A NOVEL COLD ACTIVE ESTERASE FROM THE MARINE PSYCHROTOLERANT ISOLATE *PSEUDOALTEROMONAS ARCTICA* SP. NOV.

Vom Promotionsausschuss der

Technischen Universität Hamburg-Harburg

zur Erlangung des akademischen Grades

Dr. rer. Nat.

genehmigte Dissertation

Von

M. Sc. Rami Al - Khudary

Aus Dem Libanon

Hamburg 2006

Die vorliegende Arbeit wurde im Arbeitsbereich Technische Mikrobiologie der Technischen Universität Hamburg-Harburg durchgeführt

Vorsitzender des Prüfungsausschusses: Prof. Dr.-Ing F. Keil
1. Gutachter: Prof. Dr. rer. nat. G. Antranikian
2. Gutachter: Prof. Dr. rer. nat. R. Müller
Tag der mündlichen Prüfung: 13.10.2006

Bibliografische Information Der Deutschen Bibliothek

Die Deutsche Bibliothek verzeichnet diese Publikation in der Deutschen Nationalbibliographie; detaillierte bibiografische Daten sind im Internet über http://dnb.ddb.de abrufbar.

Dissertation (Ph.D. thesis), Technische Universität Hamburg-Harburg
Printed in Germany
ISBN-10: 3-8334-6644-8
ISBN-13: 978-3-8334-6644-1

Herstellung und Verlag: Books on Demand GmbH, Norderstedt
Gedruckt auf alterungsbeständigem Papier nach ISO 9706 (säure-, holz- und chlorfrei).

Gedruckt mit Unterstützung des Deutschen Akademischen Austauschdienstes

Say: "Are those equal, those who know and those who do not know?"

Holy Quran, 039.009

To my parents Hisham and Mona, my sister Nadeen and my brother Kareem

To the soul of the late Lebanese Prime Minister Rafik Al Hariri and to his family

ABBREVIATIONS

3′	3′-end of DNA-molecule
5′	5′-end of DNA-molecule
bp	Base pairs
BSA	Bovine serum albumin
CHAPS	3-[(3-Cholamidopropyl)dimethylammonio]propanesulfonic acid
Δ	Delta, difference
Da	Dalton
dH_2O	Distilled water
DMSO	Dimethylsulfoxide
DNA	Deoxyribonucleic acid
DNase	Desoxyribonuclease
dNTP	Deoxyribonucleoside triphosphate
DTT	Dithiotreitol
E. coli	Escherichia coli
EDTA	Ethylenediamine tetraacetic acid
Est 37	*Pseudoalteromonas arctica* esterase
EtOH	Ethanol
h	Hours
IPTG	Isopropyl-β-D-thiogalactopyranoside
kbp	Kilo base pairs
kDa	Kilodalton
K_M	Michaelis-Menten-Konstante
LMW	Low molecular weight (Standard proteins)
M	Molar (mol pro Liter)
min	Minute(s)
n.d.	Not determined
OD	Optical density
OsmC	Osmotic shock response protein
ORF	Open reading frame
PAGE	Polyacrylamide gel electrophoresis
PCR	Polymerase chain reaction
pH	Negative decadic logarithm of proton concentration
pI	Isoelectric point
PMSF	Phenylmethylsulfonylfluoride
RNase	Ribonuclease

rpm	Rotations per minute	
RT	Room temperature	
SDS	Sodium dodecyl sulfate	
TEMED	N,N,N',N'-Tetramethyl-ethylendiamine	
Tm	Melting temperature	
TE	Tris-EDTA	
Tris	Tris-(hydroxymethyl)-aminomethane	
Tween-20	Polyoxyethylene sorbitan monolaurate EO 20	
Tween-80	Polyoxyethylene sorbitan monooleate EO 20	
U	Unit	
UV	Ultraviolet	
V	Volt	
Vol.	Volume	
v/v	Volume per volume	
w/v	Weight per volume	
X-Gal	5-Bromo-4-chloro-3-indoyl β-D-galactoside	

Abbreviations of amino acids

Ala	A	Alanine	Arg	R	Arginine
Asn	N	Asparagine	Asp	D	Aspartate
Cys	C	Cysteine	Glu	E	Glutamate
Gln	Q	Glutamine	Gly	G	Glycine
His	H	Histidine	Ile	I	Isoleucine
Leu	L	Leucine	Lys	K	Lysine
Met	M	Methionine	Phe	F	Phenylalanine
Pro	P	Proline	Ser	S	Serine
Thr	T	Threonine	Trp	W	Tryptophan
Tyr	Y	Tyrosine	Val	V	Valine
Xaa	X	Any Possible amino acid			

C h a p t e r   1

1 INTRODUCTION

1.1 Biocatalysis at the freezing point of water

1.1.1 Psychrophilic Microorganisms

Life under low-temperature conditions was identified as early as 1840 by Hooker, who observed that algae were associated with sea ice. In 1887, Forster was the first who reported that microorganisms isolated from fish could grow well at 0 °C (Forster, 1887). The term "psychrophilic" was first used in 1902 by Schmidt-Nielsen to describe such cold-adapted organisms (Schmidt-Nielsen, 1902). Psychrophiles are extremophilic organisms that are able to tolerate, and may even require cold conditions to survive. A distinction may be made between psychrophilic (organisms that thrive in cold environments) and psychrotolerant (organisms that thrive in mesophilic temperatures, but that can withstand cold temperatures) organisms; both terms were coined by Morita (Morita, 1975). A significant portion of the Earth is cold (e.g. Arctic, Antarctic, deep ocean), hence psychrophiles may constitute a significant portion of the living world. Psychrophiles are diverse and include members from Bacteria, Archaea, and Eukarya. A widely used definition for psychrophilic microorganisms was proposed by Morita (Morita, 1975): organisms that have optimum growth temperatures of <15° C and upper limits of ~20° C. Feller and Gerday (Feller & Gerday, 2003) identify three problems with this classic definition: (1) the temperature limits were arbitrarily determined, not based on biological processes; (2) this definition does not apply to most psychrophilic eukaryotes; (3) optimal growth temperatures are not necessarily optimal temperatures for metabolic processes. Psychrophiles' wide distribution in cold regions often requires these organisms to be tolerant to other extremes. These include environments of high

1

pressure (e.g. deep sea), high salt concentrations (e.g. sea ice), high levels of UV radiation (e.g. snow or ice cap communities), aridity (e.g. Antarctic cryptoendoliths), and low light (e.g. alpine crack or cave-dwelling communities). Recent studies (D'Amico *et al.*, 2002; Deming, 2002; Feller, 2003) are uncovering the mysteries behind the molecular adaptations that allow psychrophiles to thrive in cold conditions, however these adaptations remain poorly understood. Recent studies have also shed light on how psychrophilic metabolic activities may contribute to weathering processes and carbon/nutrient cycling (Skidmore *et al.*, 2005; Skidmore *et al.*, 2000) as well as how they may be utilized for biotechnological purposes such as bioremediation in cold regions (Cavicchioli *et al.*, 2002).

1.1.1.1   Habitats

Psychrophilic organisms are quite diverse, with members from Eukarya, Archaea, and Bacteria. Their diversity extends at the community and ecosystem level, as these communities are often isolated. While they may rely on meltwater and/or sediment influx for replenishing nutrients, cold-active microbial communities can be self-sufficient in nutrient cycling. Although some diversity in organisms exists between different cold environments, many are common throughout cold regions. Select regions will be discussed below.

Glaciers

Foght, et al. (Foght *et al.*, 2004) describe such a diverse, self-sufficient community that exists in sediments beneath two southern hemisphere glaciers (Frans Josef and Fox Glaciers, New Zealand). These communities include several ecotypes (that have been cultured), including aerobic and microaerophillic heterotrophs, N-fixing and reducing bacteria, and Ferric iron reducers. Skidmore, et al. (Skidmore *et al.*, 2005; Skidmore *et al.*, 2000) cultured similar communities from the basal (bottom, sediment-rich ice) ice of high Arctic glaciers. While the

2

organisms that comprise these glacial communities depend on the physical (e.g. temperature, light) and chemical (e.g. nutrient and oxygen content) properties of the environment, they have also been speculated to play a role in modifying their environment. Several studies suggest that the metabolic waste products from these organisms may chemically-mediate weathering of the bedrock (Foght *et al.*, 2004; Skidmore *et al.*, 2005; Skidmore *et al.*, 2000). They may also have an impact on the global Carbon cycle, as they represent a previously unconsidered population of biota that utilize Carbon (previously unconsidered because they were not known to exist).

Shain, et al. (Shain *et al.*, 2001) provide evidence of Eukaryotic life on Alaskan glaciers. They found that the 'ice worm' (*Mesenchytraeus solifugus*) inhabits the Byron Glacier, AK, among other cold environments (e.g. snow fields and avalanche cones). These metazoans are light sensitive, with activities concentrated during the nighttime/darkest hours, and have been shown to prefer living in temperatures at or near 0° C. In a later study, Napolitano and Shain (Napolitano & Shain, 2004) found these and other organisms in the phyla of Eubacteria, Protista, and Fungi at the same glacier.

Christner et al. (Christner *et al.*, 2003) also found life in isolated communities on glacier surfaces. His study focused on communities existing in cryoconite holes on the Canada Glacier in Antarctica. Cloning and sequencing organisms from these holes showed further diversity within the metazoan lineage. These metazoans included nematodes, tardigrades (small, segmented organisms similar to Arthropods), and rotifers.

Ice Sheet

While relatively scarce in the literature, there have been some studies regarding life in ice sheets, such as those in Greenland and Antarctica. These organisms may have implications in altering paleoclimate reconstructions (climate change data) by changing the gas content within ice. Microorganisms have been found to exist within ice cores taken from Lake Vostok, Antarctica, accreted subglacial ice (Priscu *et al.*, 1999) and the basal ice of a core from the Greenland Ice Sheet Project 2 (GISP2) (Miteva *et al.*, 2004). Based on these studies, the organisms cultured from these environments are similar to those found in glacial (and other cold) environments (e.g. *Sphingomonas*, *Methylobacterium*).

Ice Cap

Willerslev et al. (Willerslev *et al.*, 1999) conducted a study on the biotic remains on polar ice caps, specifically the Hans Tausen ice cap in Greenland. Willerslev took an ice core from the ice cap, melted sections in sterile containers, extracted DNA from each section and PCR amplified it, sequenced the extracted DNA, searched the GenBank database for matches and performed a BLAST search to determine what organisms were present. The search revealed a diversity of organisms such as eukaryotic fungi, plants, algae, and bacteria.

Sea Ice

According to Staley and Gosink (Staley & Gosink, 1999), microorganisms found in sea ice are composed of four major phylogenetic groups: *Cytophaga-Flavobacterium-Bacteroides* (CFB) (gram negative) group, *Proteobacteria* (including α, β, and γ members), and low and high mole percent G + C gram positive bacteria. Staley and Gosink compared members from the α –Proteobacteria (*Octadecabacter* group) γ-Proteobacteria (*Iceobacter* group) and CFB (*Polaribacter* group) at the North and South Poles to determine if these groups were cosmopolitan (distributed worldwide) or endemic (specific to one area).

While their work requires further study and characterization, they found that these organisms are similar at each pole, but that their distribution between poles is hard to explain. They suggest that if they are cosmopolitan, these strains may have been distributed between poles via deep ocean currents—these currents would have kept the organisms in a cold environment even when passing through equatorial regions.

1.1.1.2   *Phylogenetic Diversity*

The diversity of organisms can vary greatly within the different ecosystems. Thus, organisms found in a polar ice cap may differ from those found in glacial lakes, sea ice, and rocks in Antarctic dry deserts. Microbes in these environments range from piezo-psychrophiles or baro-psychrophiles (high pressure tolerant) to troglo-psychrophiles (able to evolve in the absence of light), illustrating the adaptive behavior of the psychrophiles (Feller & Gerday, 2003). Figure 1 and Table 1 illustrate the diversity of psychrophillic organisms determined via molecular analyses and placed on evolutionary trees. These trees, derived from polar sea ice and cryoconite hole communities, include members from Proteobacteria (α, β, and γ), Cyanobacteria, CFBs, Metazoa, Fungi, and Archaea.

Table 1 Phylogenetic analysis of bacterial 16S rDNA sequences from cryoconite sediment from Canada Glacier, Antarctica. GenBank ascension numbers are listed in parentheses (Christner *et al.*, 2003).

Isolate Designation	GenBankaccession #	Sequence alignment % identity	Nearest phylogenetic neighbor
CD12	AF479324	97.1	*Pseudomonas saccharophila* (AB021407)
		96.5	*Matsuebacter chitosanotabidus* (AB006851)
CD89	AF479326	97.4	*Janthinobacterium lividum* (Y08846)
		97.2	*Pseudomonas mephitica* (AB021388)
CS57	AY124340	95.6	Soil clone (AF423292)
		93.3	*Flavobacterium ferrugineum* (M28237)
CD14	AF479331	95.4	*Haloanella gallinarum* (AB035150)
		94.3	*Chryseobacterium balustinum* (M58771)
CS910	AY124339	96.0	*Flavobacterium* sp. from sea ice (U85888)
		95.8	*Flavobacterium hydatis* (M58764)
CS112	AY124338	96.3	*Flavobacterium succinicans* (D12673)
		96.3	*Flavobacterium hydatis* (M58764)
CD1	AF479325	96.4	*Arthrobacter* sp. from sea ice (AF041789)
		96.0	*Arthrobacter agilis* (X80748)
CD7	AF479339	98.8	Glacial iceisolate G200-A1 (AF479340)
		98.7	*Arthrobacter* sp. (AB039736)
CS117	AY124341	99.5	*Cryobacterium psychrophilum* (AJ297438)
		97.8	*Clavibacter michiganensis* (U30254)

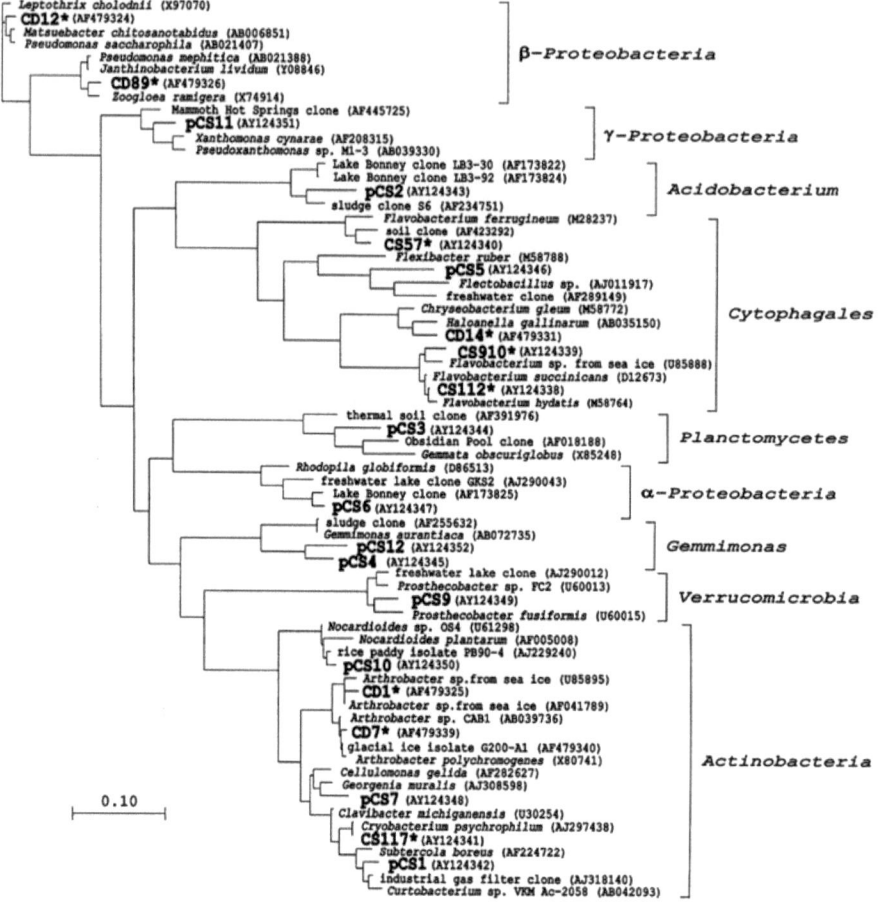

Fig. 1 Analysis of 16S rDNA amplified, cloned, and sequenced from cryoconite sediment, Canada Glacier, Antarctica. (Christner *et al.*, 2003)

1.1.2   Cold active enzymes

Psychrophiles have developed mechanisms of adaptation to temperature including a huge range of structural and physiological adjustments in order to cope with the deleterious effect of low temperatures. Indeed, they display metabolic fluxes at low temperatures that are more or less comparable to those exhibited by closely related mesophiles living at moderate temperatures (Morita, 1975). This is explained by the capability of these psychrophilic organisms to produce "cold-adapted" enzymes which are able to cope with the reduction of chemical reaction rates induced by low temperatures. However, most cellular adaptations to low temperatures and the underlying molecular mechanisms are not fully understood and are still being investigated.. These enzymes, so called "cold-active enzymes," show higher catalytic activity at low and moderate temperatures, and lower thermostability than enzymes from mesophilic and thermophilic organisms (Arpigny *et al.*, 1993; Davail *et al.*, 1994; Rentier-Delrue *et al.*, 1993). Cold-active enzymes have generated considerable interest, since they have potential as to improvement of the efficiency of industrial processes (Cavicchioli *et al.*, 2002; Gerday *et al.*, 2000). Accumulation of fundamental information on the relationship between structure and function will facilitate application of these cold-active enzymes. The enzymatical characteristics of cold-active enzymes may be attributed to their flexible structures (Feller *et al.*, 1989). The three-dimensional structures of several cold-active enzymes have been determined using the molecular modeling method (Arpigny *et al.*, 1997; Feller *et al.*, 1994; Narinx *et al.*, 1997) and X-ray crystallography (Aghajari *et al.*, 1998; Russell *et al.*, 1998), and the structural features leading to the high flexibility have been speculated to be as follows: (i) reduced numbers of hydrogen bonds, salt bridges, isoleucine clusters and proline residues in loop regions; (ii) a low Arg/(Arg + Lys) content; and (iii) an increase in the number of glycine and serine residues close to their catalytic sites. Thermodynamic studies on thermal unfolding of the coldactive phosphoglycerate kinase of a psychrophile, *Pseudomonas* sp. TACII18, revealed that the cold-active enzyme is composed of heat-labile and heat-stable domains, these domains being involved in catalytic reactions and substrate-binding of this enzyme, respectively (Bentahir *et al.*, 2000). Based on this observation, Lonhienne et al. proposed the notion of "local flexibility/rigidity," i.e. that acquisition of the flexible regions essential for the catalytic reaction and the rigid regions is important for increases in k(cat) at low temperature (Lonhienne *et al.*, 2000). The detailed structural features leading to high catalytic activity at low temperature remained to be determined. Some cold-active enzymes are shown in Table 2.

Table 2 A list of selected cold-active enzymes from psychrophilic microorganisms (Gianese *et al.*, 2001)

No.	Family name	Species	Optimal growth temperature (°C)	Databank code or reference
1	Alanine racemase	*Bacillus psychrosaccharolyticus*	15	embl:abO21683
2	α-Amylase	*Alteromonas haloplanctis* A23	4	sw:amy_altha
3	Aspartate carbamoyltransferase (catalytic chain)	*Vibrio* sp. 2693	6	sw:pyrb_vibs2
4	Aspartate carbamoyltransferase (regulatory chain)	*Vibrio* sp. 2693	6	sw:pyri- vibs2
5	ß-Galactosidase	*Arthrobacter* sp. B7	15	sw:bgal_artsp
6	ß-Lactamase	*Psychrobacter immobilis* A5	4	sw:ampc_psyim
7	Chymotrypsin A	*Gadus morhua*	4	sw:ctra~admo
8	Citrate synthase	*Antarctic bacterium* DS2-3R	5	sw:cisy _abds2
9	DNA ligase	*Pseudoalteromonas haloplanktis*	4	embl:af126866
10	Enastase	*Salmo salar*	4	nr13d:1elt
11	Isocitrate dehydrogenase 1	*Vibrio* sp. ABE-I	4	sw:idhCvibal
12	3-Isopropylmalate dehydrogenase	*Vibrio* sp. 15	15	(Wallon *et al.*, 1997)
13	L-Lactate dehydrogenase P	*Bacillus psychrosaccharolyticus*	15	sw:ldhp_bacps
14	Malate dehydrogenase	*Aquaspirillium arcticum*	4	sw:mdh_aquar
15	Ornithine carbamoyltransferase	*Vibrio* sp. 2693	6	sw:otca- vibs2
16	Pyruvate kinase	*Bacillus psychrophilus*	15	sw:kpyk_bacpy
17	Serralysin (alkaline protease)	*Pseudomonas aeruginosa* TACIII8	4	embl:psy17314
18	Subtilisin	*Bacillus* sp. TA39	4	sw:subCbacs9
19	Triosephosphate isomerase	*Vibrio marinus*	15	sw:tpis- vibma
20	Trypsin I	*Salmo salar*	4	sw:try Csalsa
21	Xylanase	*Cryptococcus adeliae* TAE85	4	embl:cay15434

1.1.3   Potential for biotechnological applications

One of the main interest of cold-active enzymes in biotechnological applications is the lack of requirement for expensive heating steps and consecutive economic benefits through energy saving. Other economic benefits originate from their abilities to function in a cold environment, to increase the reaction yield, to accommodate a high level of stereospecificity, and to minimize undesirable chemical reactions that can occur at higher temperatures (Cavicchioli *et al.*, 2002). Moreover, they prevent the requirements of an increase in enzyme concentration to compensate for the lower efficiancy when using mesophilic enzymes at low temperatures. These enzymes are also characterized by their thermal lability which allows an easy and rapid inactivation when required. This rapid inactivation of cold-active enzymes by mild heat treatment preserves product quality, permits selective enzyme inactivation in a complex medium, and does not require expensive heating/cooling systems (Margesin, 2002). The ability to heat-inactivate cold-active enzymes is particularly useful in food industry where it is important to prevent any modifications of the original heat-sensitive substrates and products. This is also useful in sequential processes such as molecular biology where the actions of an enzyme must be terminated before the next process is undertaken (Cavicchioli *et al.*, 2002). The biotechnological potential of cold - active enzymes is enhanced by the demonstration that recombinant cold-active enzymes are undistinguishable from their wild-type parent molecules with regards to kinetic parameters, folding properties (Feller *et al.*, 1998) and three-dimensional structures (Aghajari *et al.*, 1998), when the enzymes are expressed at sufficiently low temperatures (15–18°C). Moreover, expression at low temperatures is now possible as shown by the construction of a host-vector system that allows the overexpression of genes in psychrophilic bacteria (Tutino *et al.*, 2001). This study revealed a new biotechnological potential of psychrophilic strains as the expression at low temperatures prevents the formation of inclusion bodies and protects heat-sensitive gene products. Other host-vector systems for a temperature-regulated gene expression, allowing the overproduction of thermolabile enzymes originating from psychrophiles, have also been patented (Schweder, T., patent DE10101266, 2002; Kann, T. and Schweder, T., patent EP1224307, 2002). In other respects, the improvement of the competencies of psychrophilic enzymes, according to the requirement of a biotechnological process, could be obtained by directed mutagenesis or perhaps more quickly by the directed evolution of the recombinant enzyme. Possible applications of cold-adapted enzymes can be found in numerous biotechnological and industrial fields as tools in molecular biology, food, drinks, textile and detergent industries, and bioremediation (Table 3) (Cavicchioli *et al.*, 2002; Feller *et al.*, 1992; Margesin, 2002). However, despite a large number of possible biotechnological applications, only a few cold-adapted enzymes are in commercial use. To date The costs of production and processing at low temperatures remain probably higher than those for the commercial enzymes that are presently in use. Nevertheless, patent protection for enzymes from psychrophilic organisms is slowly increasing (Hasan, A.K.M.Q. and Tamiya, E., patent WO9730172, 1997; Hasan, A.K.M.Q. and Tamiya, E., patent US6200793, 1998; Kubota, H. et al., patent WO9743406, 1996; François, J.M. et al., patent WO0104276, 2001; Genot, B. et al., patent WO2004023879, 2004).

Table 3 Biochemical properties and biotechnological applications of cold-active enzymes.

Enzyme	Organism	k $_{cat}$ or V $_{max}$ (temperature)*	K$_m$ [Substrate]	T$_{opt}$ ΔH	Thermostability		Applications
					Half-Life	T$_m$†	
Metalloprotease	*Sphingomonas paucimobilis* Antarctic species	30% of V max *at −10°C*	–	30°C (15 min) 20°C (60 min) 24 kJ/mol	–	35°C (15 min)	Food, detergents, molecular biology
Serine peptidase	PA-43 subarctic bacterium	27,840/min (25°C)	3.2 mM [Amidase]	58°C	42 min (55°C) 4 min (60°C)	–	Food, detergents, molecular biology
Lipase	*Aspergillus nidulans* WG 312	*29,640/min* (40°C) 8829/min (0°C)	0.28 mM (p-[Nitrophenyl palmitate]	40°C	–	46°C (1 h)	Food, detergents, cosmetics
Alkaline phosphatase	*Vibrio sp.* G15-21	6420/min (15°C)	0.1 mM [Nitrophenyl phosphate]	–	6 min (40°C)	–	Molecular biology

Table continued on next page

Table 3 Continued

Enzyme	Organism	k cat or V max (temperature)*	Km [Substrate]	Topt ΔH	Thermostability Half-Life	Tm†	Applications
Alcohol dehydrogenase	Moraxella sp. TAE 123	550/min (25°C)	0.58 mM [Ethanol]	25°C	100% activity lost at 50°C in 30 min	-	Asymmetric chemical synthesis
3-Isopropylmalate dehydrogenase	Vibrio sp. 15	6 × 107/min (5°C)	–	48°C	10 min (60°C)	63.5°C (CD) 60°C (10 min)	Asymmetric chemical synthesis
Lactate dehydrogenase	Ice fish heart	10,200/min (3°C) 13,000/min (15°C)	16 mM (3°C) 10 mM (15°C) [Lactate]	–	–	10°C [1M Urea]	Biotransformation, biosensor, lactose removal from milk
Valine dehydrogenase	Cytophaga sp. KUC-1	2550/min (10°C)	6.8 mM [Valine]	20°C 24 kJ/mol	19 min (40°C) 2.4 min (45°C)	–	Biotransformation
β-Galactosidase	Carnobacterium piscicola BA	35,280/min (30°C)	1.7 mM [Nitrophenyl galacto-pyranoside]	30°C	85% activity lost at 20°C in 60 min	–	Dairy industries (e.g. improving quality of ice-cream and whey)
RNA polymerase	Pseudomonas syringae	3.8 nmol/mg (35°C) 0.6 nmol/mg (0°C)	–	35°C	30 min (45°C)	–	Molecular biology
DNA polymerase	Cenarchaeum symbiosum	–	–	40°C	10 min (46°C)	–	Molecular biology
DNA ligase	Pseudoalteromonas haloplanktis	0.6/min (4°C) 2.02/min (18°C)	0.17 µM (4°C) 0.3 µM (18°C) [17-mer nicked DNA]	–	24 min (18°C) 12 min (25°C)	–	Molecular biology
Uracil-DNA glycosylase	Gadus morhua	380 U/mg (37°C)	–	41°C	0.5 min (50°C) 2 h (4°C)	–	Molecular biology
Restriction endonuclease UnbI	Antarctic bacterium	–	5'-GGNCC-3' Sticky ends	15°C	–	–	Molecular biology
Triose phosphate isomerase	Vibrio marinus	4.2 × 105/min or (10°C)	1.9 mM [G3P]	–	10 min (25°C) 58 min (10°C)	41°C (DSC)	Biotransformation

Table continued on next page

11

Table 3 Continued

Enzyme	Organism	k_{cat} or V_{max} (temperature)*	K_m [Substrate]	T_{opt} ΔH	Thermostability Half-Life	T_m †	Applications
Chitobiase	Arthrobacter sp. TAD20	2400/min (5°C) 5880/min (15°C) 13,500/min (30°C)	23 µM (7°C) 33 µM (20°C) [Nitrophenyl-acetyl glucosamine]	44.7 kJ/mol	15 min (40°C)	45°C and 60°C (DSC)	Food, health products
Chitinase A	Arthrobacter sp. TAD20	102/min (15°C)	–	60 kJ/mol	20 min (50°C)	54.3°C (DSC)	Food, health products
Cellulase	Fibrobacter succinogenes S85	1200/min (4°C)	6.8 mg/ml [Carboxy-methyl cellulose]	24°C 4 kJ/mol	20 min (43°C)	–	Animal feed, textiles, detergents
Polygalacturonase (pectinase)	Sclerotinia borealis	700 U/mg (5°C) 2400 U/mg (40°C)	[Pectin]	45°C	100% activity lost at 50°C in 30 min	60°C (20 min)	Cheese ripening, fruit juice and wine industry
Pectate lyase	Pseudoalteromonas haloplanktis ANT/505	-	5g/L [Citrus pectin]	30°C	2 min (40°C)	-	Cheese ripening, fruit juice and wine industry
Nitrile hydatase	Rhodococcus sp. N-774	25% acrylamide formation at 0°C (20h)	-	35°C	100% activity – lost at 50°C in 5 min	-	Low- Temperature acrylamide synthesis
Pullulanase	Micrococcus sp.	-	0.018% [Pullulan]	50°C	-	45°C (30min)	Pullulan hydrolysis
Xylanase	Cryptococcus adeliae	888/min (5°C)	2.5 mg/ml [xylan]	45kj/mol	60 min (30°C)	48°C (DSC)	Dough fermentation, protoplast formation, wine and juice industry
Alanine racemase	Pseudomonas fluorescens TM5-2	2400U/mg (30°C) 600U/mg (0°C)	19 mM [Alanine] (30°C)	32°C 26.8 kJ/mol	-	34°C (1h)	Food storage, antibacterial agent
α-Amylase	Alteromonas haloplanktis	29,400/min (4°C)	1.1 g/L [Starch]	27°C	10 min (50°C)	45°C (intrinsic fluoresc-ence)	Detergents, dough fermentation, desizing denim jeans, pulp bleach
Glucoamylase	Candida antarctica CBS 6678	7740/min (40°C)	0.97 g/L [Starch]	57°C 59 and 33 kJ/mol	-	-	Starch hydrolysis
β-Lactamase	Psychrobacter immobilis	84,420/min (30°C)	51µM (Nitrocefin)	35°C	4 min (50°C)	-	Antibiotic degradation

Table continued on next page

Table 3 Continued

Enzyme	Organism	k cat or V max (temperature)*	Km [Substrate]	Topt ΔH	Thermostability Half-Life	Tm†	Applications
Phosphoglycerate kinase	*Pseudoalteromonas* sp. TAC 1118	30,000/min (25°C)	0.21 mM [ATP]	35°C	18 min (50°C)	51.1 °C (DSC)	Biotransformation
Catalase	*Vibrio rumoiensis*	4100U/mg (30°C)	[H₂O₂]	30°C	70% activity lost at 50°C in 60 min	-	Dairy, water treatment in paper, food, textile, semiconductor industries
Aspartate carbamoyl-transferase	*Vibrio* strain 2693	6.5 µmol/h (30°C) 1.3 µmol/h (2°C)	0.3 mM [Aspartate]	32°C 42 kJ/mol	4 min (60°C) 8 min (50°C)	50°C (15 min)	Biotransformation
Chlamysin (lysozyme-like)	*Chlamys islandica*	5x10⁵ U/mg (24°C)	[*Micrococcus luteus* cells]	24°C	20% activity lost at 23°C in 1 month	-	Antibacterial agent, food preservation
Isocitrate lyase	*Colwellia maris*	1086/min (20°C) 550/min (10°C)	510 µM (20°C) [Isocitrate]	20°C 40kJ/mol	1 min (30°C)	-	Biotransformation
Malate synthase	*Colwellia maris*	3640/min (20°C) 1100/min (10°C)	20 µM (45°C) [Glycoxylate]	45°C 28kJ/mol	1min (45°C)	-	Biotransformation

*Temperature at which *k*cat or *V* max was determined. †T m (melting temperature) determined by reaction assay (min), circular dichroism (CD), differential scanning calorimetry (DSC) or intrinsic fluorescence. G3P, glyceraldehyde-3-phosphate. (Cavicchioli *et al.*, 2002)

1.2   Esterases

Esterases and lipases are currently among the most important groups of biocatalysts in biotechnology. Throughout all species of the living world, a variety of esterolytic enzymes can be found, indicating that these enzymes were developed to perform specific metabolic reactions (Jaeger *et al.*, 1994; Reetz & Jaeger, 1998). Many of them show a wide substrate range which led to the assumption that they have evolved to enable access to carbon sources or to be involved in catabolic pathways (Bornscheuer, 2002). Bacteria produce different classes of lipolytic enzyme, including carboxylesterases (EC 3.1.1.1), which hydrolyze small ester-containing molecules at least partly soluble in water, true lipases (EC 3.1.1.3), which display maximal activity towards water-insoluble long-chain triglycerides, and various types of phospholipase (Arpigny & Jaeger, 1999).

The esterases and lipases were amongst the first enzymes tested and found to be relatively stable in organic solvents (Schmid & Verger, 1998).

1.2.1   Classification

Hydrolases form a class of enzymes that shows very wide substrate specificity. They can hydrolyze peptides, amides and halides in addition to esters and triglycerides. The fact that enzymes which show esterase activity are also able to hydrolyze non-ester bonds raises interesting questions on the terminology and classification of these enzymes (Junge & Krisch, 1973). Historically these enzymes have been classified according to their known substrate specificity. Esterases were defined as enzymes which hydrolyze ester linkages by the addition of a water molecule (Fig. 2).

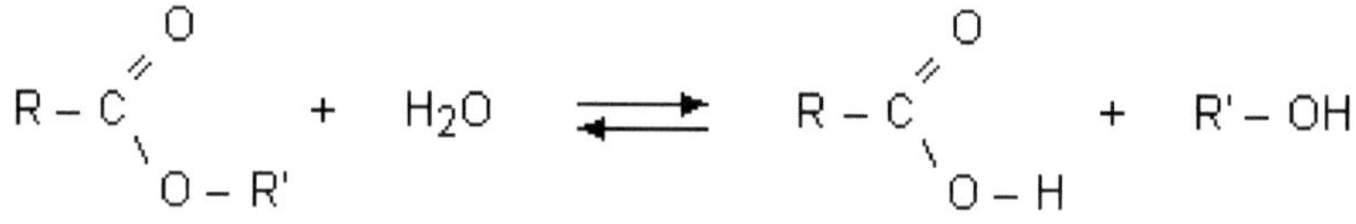

$$R - C\overset{O}{\underset{O-R'}{\big\backslash}} \quad + \quad H_2O \quad \rightleftharpoons \quad R - C\overset{O}{\underset{O-H}{\big\backslash}} \quad + \quad R'-OH$$

Fig. 2 Esterase catalyzed reaction

In general an esterase is specific for either the alcohol or the acid moiety of the substrate, but not for both. A classification scheme for esterases was proposed in 1972 by Whitaker (Whitaker, 1972), based on their specificity for the acid moiety of the substrate, such as the carboxylic ester hydrolases. The latter catalyzes the hydrolysis of carboxylic acid esters. In addition to the mentioned carboxyl esterases, aryl esterases, acetyl esterases, cholin esterases and cholesterol esterases, lipases also belong to this group of hydrolytic enzymes. For classification of these enzymes, either substrate specificity or place of enzymatic action has been used (Carriere et al., 1998). Lipases, which have a hydrophobic domain covering the active site, prefer triglycerides of long chain fatty acids, and thus have different properties than esterases, which have an acyl binding pocket (Pleiss *et al.*, 1998). Whereas esterases preferentially break ester bonds of shorter chain fatty acids, lipases display a much broader substrate range than the esterases. It appears that the physical state of the substrate is most likely a contributing factor towards the substrate specificity. Esterases preferentially hydrolyze 'simple' esters (e.g. ethyl acetate) and usually only triglycerides bearing fatty acids shorter than C6, whereas lipases prefer water-insoluble substrates, typically triglycerides composed of long-chain fatty acids (Table 4). Both enzymes have been shown to be stable and active in organic solvents, but this feature is more pronounced with lipases (Bornscheuer, 2002). Furthermore, esterases can be distinguished from lipases by the phenomenon of interfacial activation, which was only observed for lipases. Whereas esterases obey classical Michaelis-Menten kinetics, lipases need a minimum substrate concentration before high activity is observed (Fig. 3).

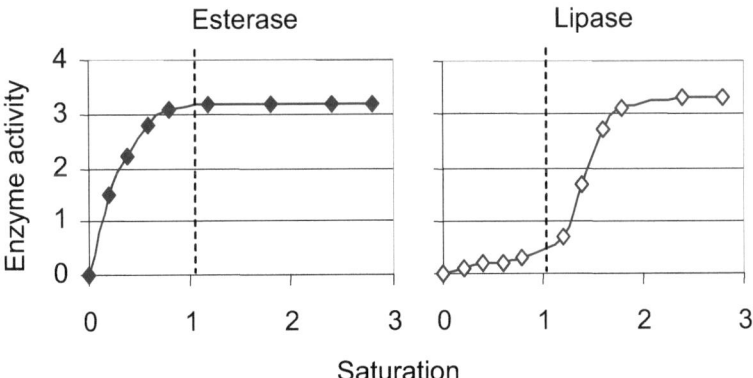

Fig. 3 Boundary surface activation of an esterase (◆) compared with that of a lipase (◇). The respective enzyme activity is a function of the substrate concentration. The broken line indicates the substrate saturation point.

Structure elucidation revealed that this interfacial activation is due to a hydrophobic domain (lid) covering the active site of lipases only in the presence of a minimum substrate concentration, i.e. a triglyceride phase or a hydrophobic organic solvent, the lid moves apart, making the active site accessible. (Bornscheuer, 2002)

Table 4 Differences between lipases and carboxyl esterases (Bornscheuer, 2002)

Property	Lipase	Esterase
Preferred substrates	Triglycerides (long-chain), secondary alcohols	Simple esters, triglycerides (short-chain)
Interfacial activation/lid	Yes	No
Substrate hydrophobicity	High	High to low
Enantioselectivity	Usually high	High to low to zero
Solvent stability	High	High to low

1.2.2   Esterase Structure

The growing knowledge of protein 3D-structures has prompted an attempt to classify proteins according to their fold. Hydrolases are found in the α/β fold group, also called α/β hydrolase fold (Schrag & Cygler, 1997). The "α/β hydrolase" superfamily proteins are primarily characterized by their common fold (α/β hydrolase fold), which is central, predominantly parallel β-sheet flanked by α-helical connections (Fig. 4). The majority of the esterases as well as all lipases share this fold. The esterases are split into three groups, the cutinase group, belonging to the flavodoxin like fold group and the esterase and acetylcholineesterase group, both having an α/β hydrolase fold. The triacylglycerole lipases contain all fungal, bacterial and pancreatic lipases.

16

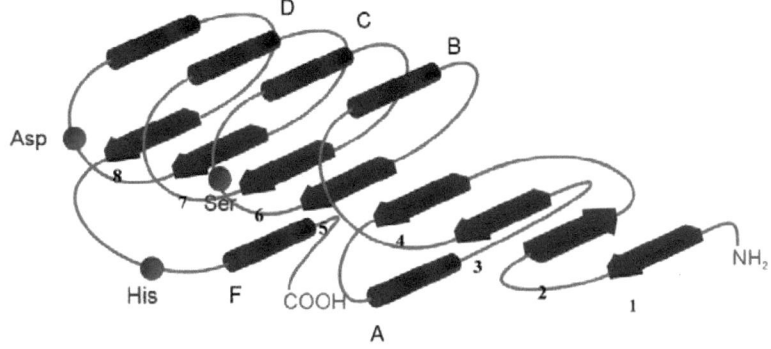

Fig. 4 Schematic presentation of the α/β-hydrolase fold. β-Sheets (1-8) are shown as arrows, α-helices (A-F) as columns. The relative positions of the amino acids of the catalytic triad are indicated as circles. (Bornscheuer, 2002)

1.2.3   Sequence motif

Esterases and lipases generally display broad substrate specificity. Thus they can not be classified solely by their function. The growing sequence information of cloned esterases and lipases has been used to identify possible sequence specific motifs. The members of the "α/β hydrolase" superfamily share a characteristic sequence motif, -Gly-Xaa-Ser-Xaa-Gly-, for most esterases and lipases, called the "nucleophilic elbow" (Nardini & Dijkstra, 1999; Ollis *et al.*, 1992). The serine residue in this motif constitutes a "catalytic triad" with Asp and His residues that are placed in this specific order (i.e., serine-aspartic acid-histidine) in the polypeptide chain. Recently, several other sequence motifs have also been found with this superfamily (Shaw *et al.*, 2002). The serine is embedded in the conserved sequence $G-X1-S-X2-G$, and ester hydrolysis is mediated by a nucleophilic attack of the active serine on the carbonyl of the substrate in a charge-relay system with the two other amino acid residues (Ollis *et al.*, 1992).

Most esterases display the conserved sequence motif (GESAG) around the central active site serine residue (Drablos & Petersen, 1997). This motif can be used as an indicator for an esterase classification of an amino acid sequence. The GESAG motif, however, is also found in the exfoliative toxin B (Vath *et al.*, 1999), which is a chymotrypsin-like protease. The GESAG motif is also found in some lipases, as examples we mention here, the lipases from *Candida rugosa, Geotrichum candidum* and *Yarrowia lipolytica*. The ProSite database (Hofmann *et al.*, 1999) classification of the active site motif for esterases is F-[GR]-G-x(4)-[LIVM]-x-[LIV]-x-G-x-S- [STAG]-G (S is the active site residue) (Cygler *et al.*, 1993). Whereas the lipase family, according to ProSite, also containing the acetylcholine esterases, has a consensus motif of [LIV]- x-[LIVFY]-[LIVMST]-G-[HYWV]-S-x-G-[GSTAC] (S is the active site residue) (Chapus *et al.*, 1988). Comparison of these motifs yields that the reported motif for lipases is also contained in the esterase motif and vice versa.

Until a few years ago, it was believed that for all lipases and carboxyl esterases only the consensus sequence motif Gly-x-Ser-x-Gly (where x represents an arbitrary amino acid residue) occurs around the active site serine. Indeed, most lipases and esterases contain this motif. More recently, a thorough comparison of 53 amino acid sequences of lipases and esterases revealed that other motifs also exist (Arpigny & Jaeger, 1999). These have been discussed in detail by Arpigny and Jaeger (Arpigny & Jaeger, 1999) and only the most important findings will be reviewed here. For instance, some lipases and an esterase from *Streptomyces scabies* contain a GDSL (Gly-Asp-Ser-Leu) consensus sequence. Moreover, structure elucidation of this esterase revealed that it contains a catalytic Ser-His dyad instead of the common Ser-Asp- His triad (Wei *et al.*, 1995). The acidic side chain, which usually stabilizes the positive charge of the active site histidine residue, is replaced by the backbone carbonyl of Trp$_{315}$ located three positions upstream of the His itself. The enzyme also has an α/β-tertiary fold, which differs substantially from the α/β -hydrolase fold. Other esterases in the GDSL group include those from *Pseudomonas aeruginosa* (accession code: AF005091), *Salmonella typhimurium* (AF047014) and *Photorhabdus luminescens* (X66379), the first two being outer-membrane-bound esterases. Other enzymes show high homology to the mammalian hormone-sensitive lipase family. Here, conserved sequence blocks were found, which initially have been related to activity at low temperature. However, it was found that esterases from psychrophilic (e.g. *Moraxella* sp., X53869; *Psychrobacter immobilis*, X67712) as well as mesophilic (*Escherichia coli*, AE000153) and thermophilic (*Archeoglobus fulgidus*, AAE000985) origins belong to this family. Members of family V, such as esterases from *Sulfolobus acidocaldarius* (AF071233) and *Acetobacter pasteurianus* (AB013096), share significant homology to non-lipolytic enzymes, e.g. epoxide hydrolase, dehalogenase and haloperoxidase. Rather small (23-26 kDa) enzymes are found in family VI, which includes an esterase from *P. fluorescens*, of which the structure is known (Kim *et al.*, 1997). The esterase is active as a dimer, has a typical Ser-Asp-His catalytic triad and hydrolyzes small substrates, but not long-chain triglycerides. Interestingly, ~ 40% homology to eukaryotic lysophospholipases is found for members of this family. In contrast, esterases from family VII are rather large (~55 kDa) and share significant homology to eukaryotic acetylcholine esterases and intestine or liver carboxyl esterases (e.g. pig liver esterase). A p-nitrobenzyl esterase from *Bacillus subtilis* (Moore & Arnold, 1996; Zock *et al.*, 1994) and an esterase from *Arthrobacter oxydans* (Q01470) active against phenylcarbamate herbicides (Pohlenz *et al.*, 1992) belong to this group. In the last family, VIII, high homology to class C L-lactamases is observed. These enzymes contain a Gly-x-Ser-x-Gly motif and a Serx- x-Lys motif, but it has recently been demonstrated by site-directed mutagenesis studies of an esterase (EstB) from *Burkholderia gladioli* that the Gly-x-Ser-x-Gly motif does not play a significant role in enzyme function (Petersen *et al.*, 2001). The most prominent member is an esterase from *Arthrobacter globiformis* (AAA99492) (Nishizawa *et al.*, 1995), which stereoselectively forms an important precursor of pyrethrin insecticides.

1.2.4   Mechanism of action

The mechanism for ester hydrolysis or formation is essentially the same for lipases and esterases and is composed of four steps: First, the substrate is bound to the active serine, yielding a tetrahedral intermediate stabilized by the catalytic His and Asp residues. Next, the alcohol is released and an acyl-enzyme complex is formed. Attack of a nucleophile (water in hydrolysis, alcohol or ester in (trans-)esterification) forms again a tetrahedral intermediate, which after resolution yields the product (an acid or an ester) and free enzyme (Bornscheuer, 2002).

1.3   Microbial esterases

Preliminary experimental findings, together with knowledge of the specific microorganisms and kind of esterase required, suggest the choice of the appropriate fermentation technique for production of a specific esterase (Shankaranand *et al.*, 1992). Esterases like, ferulic acid esterases, feruloyl esterases, and cinnamic acid esterases, are produced from *Aspergillus* sp. (Faulds *et al.*, 1997; Topakas *et al.*, 2003) using either solid state or submerged fermentations (Asther *et al.*, 2002), whereas extracellular sterol esterase is produced by *Ophistoma piceae* in submerged fermentation growing on glucose and on different C-sources (like olive oil, agro-industrial by-products) (Calero-Rueda *et al.*, 2002). In most cases, such substrates contain the required inducer. On the other hand, phosphotriesterase from *Pseudomonas montelli* (Kawanami *et al.*, 2002), which is regulated by phosphate and C sources, thermostable intracellular esterase from *Bacillus* sp. (Kademi *et al.*, 2000), and tributyrin-induced esterase in *Lactobacillus casei* (Choi & Lee, 2001) differ from the enzymes produced by *Aspergillus* sp. or *Sporotrichum* sp. However, in all of these fermentations, other hydrolytic enzymes like cellulases, hemicellulases, etc., are also produced that incur severe costs in downstream separation. It is also pertinent to mention that human and pig esterases, for example, cannot be produced by fermentation of the wild type organism. Thus, recombinant production methods are employed to facilitate production of pure esterases and the molecular engineering of enzymes for specific applications.(Panda & Gowrishankar, 2005)

1.3.1   Eukaryotic esterases

The *Aspergillus niger* re-cinnamoyl esterase (reFAEA) is expressed to a high level by *Pichia pastoris* (Juge *et al.*, 2001). Studies by de Vries and Visser (de Vries & Visser, 1999) on regulation of the feruloyl esterase (faeA) gene from *Aspergillus* showed that the enzyme is synthesized separately on xylan and pectin. The gene encoding FAEB—forming a homodimer protein with FAEA—was amplified from genomic DNA of *A. niger* using an upstream primer conserving the signal sequence required for endoplasmic reticulum translocation, and a downstream primer with a six his-codon tag to facilitate efficient affinity chromatographic separation (Levasseur *et al.*, 2004). Expression of the homologous faeB gene in *A. niger* produced esterase activity by day 3, which increased to a maximum production on day 11 of 18 nkat/ml (Panda & Gowrishankar, 2005).

Esterases play a major role in the detection of traces of insecticides in the environment. The acetylcholinesterase (AchE) gene of *Drosophila* was engineered by site-directed mutagenesis to increase its sensitivity and its rate of phosphorylation or carbomylation by organophosphates and carbamates (Boublik *et al.*, 2002). For organophosphate degradation, a number of genes have been identified and studied like the in vitro modified carboxylesterase E3 from *Lucilia cuprina* with a mutation (W251G) conferring 20-fold higher malathion carboxylesterase (MCE) activity and mutations at other sites near the bottom of the catalytic cleft diminishing MCE (Heidari *et al.*, 2004) and an orthologus *Musca domestica* Md αE7 gene displaying a structure similar to that of LC αE7 having a mutation in Gly137 (Claudianos *et al.*, 1999).

Studies on mammalian carboxyl esterases show that they constitute a multigene family, being localized in the endoplasmic reticulum of many tissues. They

21

catalyze the hydrolysis of various drugs or prodrugs containing ester or amide bonds (Satoh *et al.*, 2002). *r*-DNA and expression techniques have been exploited for these enzymes (for liver, intestine and brain carboxylesterases) to elucidate catalytic mechanisms, substrate specificity, etc. The pig liver esterase gene, cloned and modified by site-directed mutagenesis, can be expressed extracellularly in *Pichia pastoris* (Musidlowska-Persson & Bornscheuer, 2003).

1.3.2   Bacterial esterases

Pseudomonas species have been studied thoroughly as they produce esterases for a variety of applications. An extracellular polyurethanase gene (*pueB*-3.2 kb) was cloned from *Pseudomonas* into *Escherichia coli*. A BLASTP search showed that PueB shares more amino acid identity with polyurethanase than with PueA of *Pseudomonas,* and also has some similarity with *Pseudomonas* lipase. BLASTP and parsimony analyses of the predicted amino acid sequences of PueB, PueA, PudA and PulA polyurethanase enzymes have also been reported (Howard *et al.*, 2001). *Pseudomonas flurorescence* esterase (PFE) expressed in *E. coli* displayed high rate and enantioselectivity in the organic phase for α-phenylthanol with vinyl acetate compared to acylation of a series of 1,2-Oprotected glycerol derivatives and hydrolysis of 3-phenylbutyric acid methyl ester (Krebsfanger *et al.*, 1998). For D-β-acetylthioisobutyrate production, *Pseudomonas putida* esterase was cloned and expressed in *E. coli* JM109 using the vector pUC19 (Ozaki *et al.*, 1994). Plasmid pPE 101, containing an 8 kb insert, was subcloned as a 1.2 kb PstI–ClaI fragment in pPE117. A cosmid library of *Pseudomonas fluorescens* has been constructed in *E. coli* HB101 using plasmid pHC79 to study p-nitrophenol degradation. Furthermore, vdh, vanA, and vanB subunit genes from *Pseudomonas* sp. have been reported for vanillin bio-conversion (Priefert *et al.*, 1997).

Thermophilic esterases are another emerging avenue. For example two carboxyl esterases of *Geobacillus stearothermophilus* (est. 55 and est. 30) have a half-life at 70°C of 40 and 180 min, respectively (Ewis *et al.*, 2004). Maximal activity was observed with C6 p-nitrophenylacylates of C4 and C6 for Est. 55 and for C6 of Est. 30. Inhibition studies showed both serine and histidine residues in their active sites. Sequence and structural homology analysis showed that both these carboxyl esterases belong to the α/β hydrolase family, with different functional amino acids at the catalytic site.

1.3.3   Archaeal esterases

A carboxylesterase from *Sulfolobus sulfataricus* (for which a gene library has been constructed in *E. coli*) is encoded on plasmid isolated from an EstA clone. Restriction analysis revealed this gene to be present on a 3.3 kb DNA chromosomal fragment and a multiple sequence alignment with enzymes from other species has been reported (Morana *et al.*, 2002). Also, a hyper-thermoactive esterase from *Aeropyrum perix* K1 [APE1547 gene encoding 582 a. a. residues cloned into *E. coli* BL21(DE3) using vector pET11a with a T7 promoter] produces an enzyme possessing a lipase motif and low homology with thermophilic esterases (Gao *et al.*, 2003). Both archaeons have been investigated with respect to thermostability of the enzyme. For *Sulfolobus* sp., studies by Suzuki et al. (Suzuki *et al.*, 2004) showed optimum conditions for ester cleavage of 70°C and pH 7.5–8.0. In this case, thermostability has been established in both aqueous and organic solvents. This can be exploited in the high temperature production of synthetic compounds, whereas the study of Gao et al. (Gao *et al.*, 2003) has yet to establish thermostability in organic solvents and substrate specificity of enzyme.

1.3.4   Cold-active esterases

The " cold activity" (i.e., high catalytic activity at low temperatures) and thermostability of esterases and lipases can be the key to the success in some of their applications (Margesin & Shinner, 1994). These applications include their use as additives in laundry detergents for cold washing and catalysts for organic syntheses of unstable compounds at low temperatures. So far, several cold adapted esterases and lipases from cold-adapted microorganisms have been studied, and their potential application have been examined (Choo *et al.*, 1998; Ferrer *et al.*, 2004; Kulakova *et al.*, 2004; Suzuki *et al.*, 2003; Suzuki *et al.*, 2001, 2002). An esterase gene - hormone-sensitive lipase group of lipase/esterase family - of *Psychrobacter* sp., isolated from Antarctic soil, was cloned, sequenced and expressed in *E. coli*. High activity was observed between 5 and 25°C with a half-life of 16 min for thermal inactivation at 40°C and pH 7–9 (Kulakova *et al.*, 2004). The activity at low temperature and thermal stability are probably due to flexibility around the active site of the enzyme. Two more cold- active esterases from *Acinetobacter* sp. (Suzuki *et al.*, 2002) and *Pseudomonas* sp. (Suzuki *et al.*, 2003) have been cloned and studied.

1.3.5   *Directed evolution and future trends*

Directed evolution has been developed since the early 1990s to obtain novel biocatalysts. This technique has been used successfully to improve enantioselectivity, thermostability, and yield of the required enzyme (Horsman *et al.*, 2003; Jaeger *et al.*, 2001; Van Kampen & Egmond, 2000). In this case, a library of mutants is developed using random mutagenesis or by gene shuffling. This library is then assayed by high-throughput screening techniques (Baumann *et al.*, 2000; Liu *et al.*, 2001; Wahler & Reymond, 2001). High-throughput enzyme assays often use novel fluorogenic and chromogenic substrates. John and Heinzle (John & Heinzle, 2000) have developed a quantitative microplate screening method for hydrolases using pH indicators. Photometric and fluorometric assays are usually performed in microtiter plates using sophisticated robotics. For successful directed evolution of esterases, effective mutation strategies, suitable expression systems, fast and sensitive assays for identification and enzyme variants are required. Stability and activity relationships of a mesophilic esterase have been examined using *in vitro* evolution. The activity of thermostable p-nitrobenzyl esterase (p-NBE) variants has been determined using a 96-well plate assay based on esterase activity. The thermal stability of an enzyme can be increased significantly without loss of activity at lower temperatures by directed evolution. Reetz (Reetz, 2000) has reviewed in detail the application of directed evolution in the development of enantioselective enzymes. The sequential improvement in enantioselectivity of enzyme- catalyzed reactions can be achieved, if an action causing evolutionary pressure is performed in reiterative cycles. Furthermore, the versatility of substrate-binding motifs suggests the selection of several family members in addition to non-catalytic signal transduction functions (Oakeshott et al., 1999).

In summary, there has been some progress in molecular biology research on esterase to express animal- and plant source genes in suitable expression systems. Detailed analysis is required on the engineering aspects of production like proper optimization of the process, reactor analysis, kinetics and modeling of fermentation systems. The advantage of one particular source for a particular application of esterase (e.g., health or environmental issues) requires further interpretation. Esterases from animal and plant origin may be designed as effective drugs or as detection probes for hazardous compounds using chiral technology (Panda & Gowrishankar, 2005).

1.4   Application of esterases

1.4.1   *Food processing, beverages, perfume industries, and degradation of synthetic materials*

Ferulic, sinapic, caffeic, and coumaric acids are widely used in the food, beverage, and perfume industries (Chaabouni et al., 1996). Their esters are present in cereals, agro-industrial residues, and biopulp (Asther et al., 2002). Feruloyl and cinnamoyl esterases of *Aspergillus niger*, along with pectinases, cellulases and xylanases, can release such hydroxyl cinnamic acids from wheat bran, rice bran, sugar cane, bamboos, sugar beet pulp, etc. An esterase from *Fusarium oxysporum* plays a significant role in producing flavoring and fragrance

25

compounds from geraniol and fatty acids (Chaabouni *et al.*, 1996; Christakopoulos *et al.*, 1998). Pentylferulate ester, a flavor precursor in food processing, and also in cosmetics, is a product of feruloyl esterase using water-in-oil microemulsions (Giuliani *et al.*, 2001). However, research is being directed towards the improvement of the flavor of fermented meat products, e.g., lipase-esterase from *Pedicoccus pentosauces* SV6 is employed for the improvement of flavor in fermented sausages (Ostdal *et al.*, 1996). Esterases are employed in dairies, and for the production of wine, fruit juices, beer, and alcohol. In order to transform low value fats and oils into more valuable ones, esterases as well as lipases are used as trans-esterification catalysts. For example, esterases and lipases from *Lactobacillus casei* CL96 are used significantly for hydrolysis of milk fat for the purpose of flavor enhancement in the manufacture of cheese-related products (Choi & Lee, 2001). A fruity strawberry aroma has been developed with esterases and lipases derived from *Pseudomonas fragi* (Kermasha *et al.*, 2000). It has also been found that an esterase from yeast plays a significant role in determining the final ester level in products such as membrane-filtered beer and bottle re-fermented beer (Dufour & Bing, 2001). Along with acetyltransferase, an esterase from *Saccharomyces cerevisiae* plays an important part in the production of isoamyl acetate, which has a major role in the determination of sake flavor (Fukuda *et al.*, 1998). These industries also use polymer-derived compounds either as packaging or as a component in their manufacture. Some man-made pollutants like plastics, polyurethane, polyesters, polyethylene glycol adipate, etc., are thus generated. To degrade such components, cholesterol esterase and polyurethanase are widely used (Jahangir *et al.*, 2003). Polyurethanases from *Pseudomonas chlororophis* and *Pseudomonas aeruginosa*, used widely in the degradation of polyester, resemble carboxylesterases and acetyl cholinesterases (Howard *et al.*, 2001).

Table 5 : Applications of esterases (Panda & Gowrishankar, 2005)

Form of esterase	Nature of application	Source	Reference
Acetylcholinesterase	Development of new drugs for schistosomiasis, biomarker for organophosphates in marine environment, assessment of poison due to pesticides and heavy metals	Blood of *Schistosoma* sp., *Mytilus edulis, Drawida willsi*	(Bentley *et al.*, 2003; Brown *et al.*, 2004; Panda & Sahu, 2004)
Acetyl esterase, methyl esterase, acetylglucomannan esterase and acetyl xylan esterase	Release of acetyl and methyl residues from cell wall, degradation of cellulose, acetic acid from O-acetyl-galactoglucomannan and O-acetyl-4-O-methyl-glucuronoxylan	*Aspergillus, Trichoderma* sp	(de Vries & Visser, 1999; Poutanen *et al.*, 1990; Puls *et al.*, 2001; Tenkanen *et al.*, 1995)
Aryl esterase	For flavor development in food and alcoholic beverages	*Saccharomyces cerevisiae*	(Lomolino *et al.*, 2003)
Carboxylesterases	Degradation of ethylene glycol dibenzoate ester, lowering toxicity of malathion, hydrolysis of aspirin and organophosphorous insecticides, determining metabolic resistances to pyretheroid insecticides, D-acetylthioisobutyric acid, synthesis of racemates of esters of 1,2-O-isopropylideneglycerol, PHA depolymerase	*Streptomyces* lividans, livers of rat and guinea pig, *Lucilia cuprina, Pediculus capitis, Bacillus coagulans*	(Biely *et al.*, 1996; Heidari *et al.*, 2004; Molinavi *et al.*, 1996; Ozaki & Sakashita, 1997; Picollo *et al.*, 2000; Riegels *et al.*, 1997; Vincent & Lagreu, 1981)
Cephalosporin acetyl esterase	Detecting acetyl groups from cephalosporin derivatives	*Burkholderia gladioli*	(Peterson *et al.*, 2001)
Cholesterol esterase and pseudocholinesterase, cholinesterase	Degradation of poly (ether-urethane), pre-requisite for working of sodium pump	Rat liver and other sources	(Jahangir *et al.*, 2003; Wheeler *et al.*, 1972)
Cinnamoyl ester hydrolase	Plant cell wall degradation	*Piromyces equi*	(Fillingham *et al.*, 1999)
Erythromycin esterase	Clinical medicine in human, poultry and fish	*Pseudomonas* sp	(Kim *et al.*, 2002a; Kim *et al.*, 2002b)
Ferulic acid esterase	Release of ferulic acid	*Aspergillus niger, Pencillium* sp	(Asther *et al.*, 2002; Kroon *et al.*, 1996)
Feruloyl esterase	Synthesis of pentylferulate ester used in cosmetics and perfumes industries, decolorization of paper mill effluent	*A. niger, Streptomyces avermitilis*	(Garcia *et al.*, 1998; Giuliani *et al.*, 2001)

Table continued on next page

Table 5 Continued

Form of esterase	Nature of application	Source	Reference
Esterases	Transesterification reactions in organic solvents, resolution of (R, S)-β-acetyl-mercaptoisobutyrate, conversion of (R,S)-ketoprofen ethyl ester and linalyl acetate, food processing and dairy industries, hydrolysis of esters of tertiary alcohol, cefditoren pivoxil (a prodrug)—(US Patent 4839350), fluorescein diacetate and 5-(6)-carboxyfluorescein diacetate to detect yeast in food, production of isoamyl acetate and mannitol, detoxification of xenobiotics, control of physiological process of hormone, hydrolysis of diphthalates, ofloxacin, microbial activity of soil, flavor quality of sake, fermented sausages, detection of methyl-parathion resistance and malathion susceptibility, improvement of aroma and flavor, fatty acid production	*Fusarium oxysporum*, recomb. *Escherichia coli*, *Pseudomonas* sp., *Burkholderia gladioli*, *Lactobacillus casei*, *Rhodococcus* sp., *Saccharomyces cerevisiae*, *Bacillus* sp., *Pediococcus pentosaceus*, *Diabrotica virgifera*, *Locusta migratoria manilensis*, *Micrococcus* sp	(Breeuwer *et al.*, 1995; Christakopoulos *et al.*, 1998; Costenoble *et al.*, 2003; Fernandez *et al.*, 2004; Fukuda *et al.*, 1998; Gokul, 1999; Gudelj *et al.*, 1998; He *et al.*, 2004; Jung *et al.*, 2003; Kermasha *et al.*, 2000; Kim *et al.*, 2002a; Kim *et al.*, 2004; Kim *et al.*, 2002b; Laranja *et al.*, 2003; Ostdal *et al.*, 1996; Peterson *et al.*, 2001; Valarini *et al.*, 2003; Wezel *et al.*, 2000; Zhou *et al.*, 2004)
Esterases from human system	Retinyl palmitate to retinol, resistant against inflammatory cells lysosomal enzymes, biodegradation of dental composites, metabolism of aspirin and non-narcotic analgesics, conversion of proparacetamol to paracetamol, hydrolysis of acetylsalicylate to salicylic acid in plasma, conversion of oseltamivir phosphate to oseltamivir carboxylate, activation of etoposide prodrugs, hydrolysis of succinylcholine and procaine		(Abbott, 2001; Chavkin, 2004; Coppens *et al.*, 2002; de Ruiter & de Haan, 2003; Finer *et al.*, 2004; Fu *et al.*, 2002; Tang *et al.*, 1997; Wrasidlo *et al.*, 2002)
Methyl jasmonate esterase	Hydrolyzing methyl esters of abscisic acid, indole-3-acetic acid and fatty acids	*Lycopersicon esculetum*	(Stuhlfelder *et al.*, 2002)

Table continued on next page

Table 5 Continued

Form of esterase	Nature of application	Source	Reference
Phosphotriesterase	Hydrolyzed product of coumaphos and coroxon	*Pseudomonas monteilli*	(Kermasha *et al.*, 2000)
Pig liver esterase, porcine liver esterase and recombinant pig liver esterase	Desymmetrization of a centrosymmetric cyclohexanediacetate, enantioselective production of levofloxacin from ofloxacin butyl ester, kinetic resolution of (R, S)-1-phenyl-3-butyl acetate and (R, S)-1-phenyl-2-pentyl acetate		(Bohm *et al.*, 2003; Choi & Lee, 2001; Musidlowska-Persson & Bornscheuer, 2003)
Polyurethanase	Degradation of polyester polyurethane and polyether polyurethane	*Comamonas acidovorans, Pseudomonas chlororaphis*	(Howard *et al.*, 2001)
Recombinant esterase (PF1-K)	Preparation of (S)-flurbiprofen	*Pseudomonas* sp	(Baron *et al.*, 1980)
Sterol esterase, steryl esterase and Cholesteryl esterase	Paper manufacturing, to reduce pitch problems during paper manufacture	*Ophiostoma piceae, Pseudomonas* sp., *Chromobacterium*	(Calero-Rueda *et al.*, 2002; Kontkanen *et al.*, 2004)

1.4.2   *Pharmaceutical industries*

Esterase plays a major role in the synthesis of chiral drugs, which are highly useful in curing certain diseases (Bornscheuer, 2002). For example, an esterase from *Trichosporon brassicae* has been used extensively for the production of optically pure (S)- and/or (R)-ketoprofen [2- (3-benzoylphenyl) propionic acid], which is very effective in the reduction of inflammation and relief of pain resulting from arthritis, sunburn, menstruation, and fever (Shen *et al.*, 2002). Esterases from *Trichosporon brassicae* and other microorganisms such as *Rhodococcus* sp., *Bacillus circulans*, etc., can also produce a large number of chiral compounds. In addition to the above strains, esterases from *Pseudomonas* sp. produce commercially available anti-inflammatory drugs (NSAIDs) such as ibuprofen [(R, S)-2- (4-isobutylphenyl) propionic acid] (Kim *et al.*, 2002a; Kim *et al.*, 2002b). Similarly, an esterase from *Bacillus coagulans* catalyzes the enantioselective hydrolysis of the racemic ester of 1,2- O-isopropylidene glycerol, a bioactive molecule (Molinavi *et al.*, 1996).

Esterase and lipases, commonly used as indicated above, were found not to be capable of catalyzing the bulky esters of tertiary alcohols. However, these can be efficiently hydrolyzed by a novel esterase (a typical carboxylesterase, EstB) from *Burkholderia gladioli*, which hydrolyzes bulky substrates such as linalyl acetate. This feature has relevance in industrial bio-catalytic applications in the production of semi-synthetic cephalosporin derivatives (Peterson *et al.*, 2001). In addition, esterase RR1 from recombinant strain *Rhodococcus* is effective in hydrolyzing the acetate ester of the allylic tertiary alcohol linalool (Gudelj *et al.*, 1998). Musidlowska-Persson and Bornscheuer (Musidlowska-Persson & Bornscheuer, 2003) have successfully demonstrated significant differences in the kinetic resolution of acetates of secondary alcohols using recombinant pig liver esterase and commercially available crude pig liver esterase. They observed that recombinant pig liver esterase possesses higher enantio-selectivity towards hydrolysis of (R, S)-1-phenyl-3-butyl acetate, (R, S)-1- phenyl-propyl acetate, (R, S)-1-phenyl-2-pentyl acetate and (R, S)-1-phenyl-2-butyl acetate. A molecular mechanism for enantio-recognition of tertiary alcohols using carboxylesterase (acetylcholinesterase and p-nitrobenzyl esterase) was reported by Henke et al. (Henke & Bornscheuer, 2002). Furthermore, 3-methoxy-1-butanol, 3-methyl-1-pentanol, 3,7-dimethyl- 1-octanol and β-citronellol are produced using pig liver carboxylesterase in aqueous medium immobilized in Sepharose or Chromosorb (Campon & Kilbanon, 1984). Stereoselective esterification of 1-trimethylsilyl-2-propanol, 1-trimethylsilyl-1-propanol and 2-methylsilyl-1-propanol could be obtained using steapsin from pig pancreas along with other microbial esterases (Uejima *et al.*, 1993). Porcine liver esterase is also effective in the enantioselective production of levofloxacin from ofloxacin butyl ester (Puls *et al.*, 2001). Patel (Patel, 2000) reviewed stereospecific conversions in the production of pharmaceutical intermediates with respect to taxol semisyntheses like throumboxane-A2-antagonist, acetylcholine esterase inhibitors, anticholesterol drugs, anti-infective drugs, Ca-channel blocking drugs, antiarrhythumic agents, K-channel openers (from *Mortierells ramanniana*) and antiviral agents. In addition, Schutt (Schutt, 1996), Kanegafuchi-chem (Kanegafuchi-Chem, 1991), Squibb (Squibb, 1993) and Duphar-Int Res (Duphar-Int-Res, 1994) reported a number of stereoselective compounds using stereo-selective esterase. Ozaki and Sakashita (Ozaki & Sakashita, 1997) produced D-p-acetylthioisobutyrate from methyl-D,L-β-acetylthioisobutyrate using a thermostable esterase of *Pseudomonas putida*. Human serum esterase (carboxylesterase) is used to hydrolyze, and to reduce the toxicity of, the prodrugs etoposide. Derivatives of etoposide are widely used as highly effective anticancer drugs against a broad spectrum of tumors, including pediatric cancers such as acute lymphatic lymphomas, rhabdomyosarcomas and neuroblastomas (Wrasidlo *et al.*, 2002).

1.4.3   *Chemical industries*

Although a considerable number of microbial carboxyl esterases are known which have been overexpressed in suitable hosts, only a few of them have been used for the synthesis of optically pure compounds. The major reasons for this are their limited commercial availability and their frequently observed moderate enantioselectivity. Several esterases have been available in recent years from various suppliers (e.g. Fluka, Amano, Jülich Fine Chemicals, Diversa, Roche Diagnostics, and Thermogen). Probably the best studied enzyme is the so-called carboxyl esterase NP (NP from naproxen, a non-steroidal anti-inflammatory drug) originating from *B. subtilis* (Quax & Broekhuizen, 1994). Besides naproxen, various other 2-arylpropionic acids are produced with high enantioselectivity (Fig. 5) (Azzolina *et al.*, 1995a; Azzolina *et al.*, 1995b). Carboxyl esterase NP has a molecular mass of 32 kDa, a pH optimum between 8.5 and 10.5 and a temperature optimum between 35 and 55³C. Carboxyl esterase NP is produced as intracellular protein; its structure is unknown. In a pilot-scale process, (*R,S*)-naproxen methylester is hydrolyzed in the presence of Tween 80 to increase substrate solubility at pH 9.0. The (*S*)-acid is separated from the remaining (*R*)-methylester, and the latter is racemized using an organic base. This reaction yields (*S*)-naproxen with excellent optical purity (99% enantiomeric excess (ee)) at an overall yield of 95% (Quax & Broekhuizen, 1994).

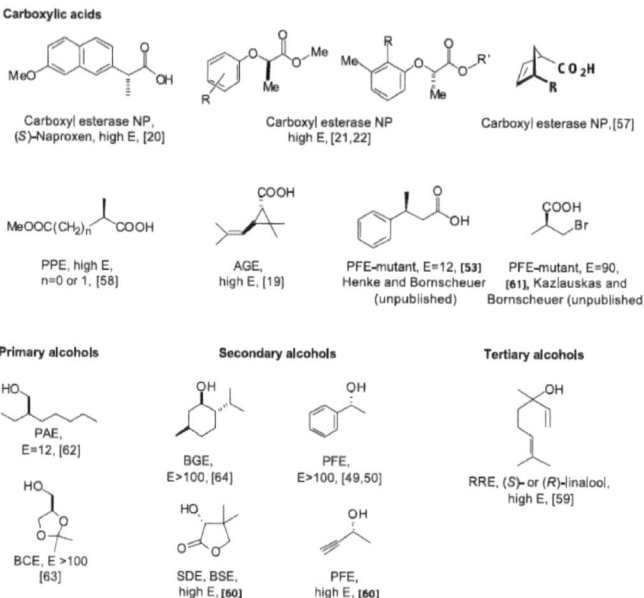

Fig. 5 Selected examples of chiral compounds obtained by carboxyl esterase-catalyzed kinetic resolutions. AGE, esterase from *A. globiformis*; BCE, esterase from *Bacillus coagulans*; BSE, esterase from *Bacillus stearothermophilus*; BGE, esterase from *B. gladioli* ; PAE, esterase from *P. aeruginosa*; PME, esterase *from Pseudomonas marginata*; PPE, esterase from *Pseudomonas putida*; RRE, esterase from *Rhodococcus ruber*; SDE, esterase from *Streptomyces diastatochromogenes*. The E value is the enantioselectivity (often also named enantiomeric ratio), which reflects the ability of the enzyme to distinguish between the two enantiomers of a racemate and thus the rate with which each enantiomer is converted. E values above 100 allow the synthesis of optically pure product and substrate, E values > 20 are sufficient to obtain the remaining substrate in high optical purity and acceptable yield. At E =1, racemic product is formed. E values can be calculated from the Vmax/Km values for each enantiomer, but are usually determined from % ee and conversion using the equations developed by Chen et al. (Chen *et al.*, 1982). A simple program to calculate the enantioselectivity is freely available at http://www.orgc.tu-graz.ac.at (Bornscheuer, 2002).

Carboxyl esterase NP was also used in the resolution of (R,S)-ibuprofen methylester and showed higher selectivity compared to lipase from *Candida rugosa* (Mustranta, 1992). Another efficient kinetic resolution was achieved in the synthesis of (+)-trans-(1R,3R)-chrysanthemic acid, which is an important precursor of pyrethrin insecticides (Fig. 5). Here, an esterase from *A. globiformis* catalyzed the sole formation of the desired enantiomer (>99% ee, at 77% conversion). The enzyme was purified and the gene was cloned in *E. coli* (Nishizawa *et al.*, 1993). Hydrolysis was performed at pH 9.5 and 50°C. Acid produced was separated through a hollow-fiber membrane module and the esterase was very stable over four cycles of 48 h (Nishizawa *et al.*, 1995). Further selected examples for the application of microbial carboxyl esterases in the synthesis of optically pure compounds are summarized in (Fig. 5).

1.4.4   *Other applications of esterases*

Esterase is extensively used in other industries, such as the pulp and paper, textile, leather and baking industries. Sterol esterase from *Ophiostoma piceae* is applied in paper manufacture, as it efficiently hydrolyzes both triglycerides and sterol esters. Further, steryl esterase and cholesteryl esterase from *Pseudomonas* sp., *Chromobacterium viscosum* and *Candida rugosa* also play a significant role in reducing pitch problems during paper manufacture. Pitch deposits, which have a negative impact on paper machine runability and paper quality, are formed during the manufacture of softwood and hardwood paper pulps (Kontkanen *et al.*, 2004). Phosphotriesters, e.g., synthetic organophosphorous compounds such as coumaphos, and its oxon derivative, coroxon, have broad applications as insecticides and nematicides. The residues of these compounds cause toxicity towards the environment and can enter foodstuffs. In this context, phosphotriesterases from *Brevundimonas diminuta* and *Alteromonas* sp. have been used extensively in detoxifying/degrading these organophosphorous compounds (Horne *et al.*, 2002).

Other applications for esterases also exist and, in the future, novel applications in acute areas such as health and environmental aspects may be predicted. However, to efficiently exploit specific application of esterase, particular esterases need to be produced efficiently and thoroughly characterized (Panda & Gowrishankar, 2005).

1.5   Aim of the study

Most habitats on our planet are permanently cold. By volume, 90% of the world's oceans have a temperature of 5°C or less, supporting both psychrophilic and psychrotolerant microorganisms. When terrestrial habitats are included, over 80% of the earth's biosphere is permanently cold (Russell, 1990). Researchers have reported on cold-adapted microbes as early as 1887. However, within the last five years, the number of papers dealing with cold adaptation has increased sharply. This is true both for the analysis of the cold shock response as well as the description of new psychrophilic and psychrotolerant species, especially from permanently cold habitats. This sharp increase in interest is certainly fueled by many emerging biotechnological uses of cold tolerant organisms.

In recent years, growing attention has been devoted to cold adapted microorganisms and their enzymes. The number of available enzymes that have great potential for industrial chemical reactions has increased considerably since the 1980s, mainly as a result of enormous achievements in the cloning and expression of enzymes from diverse culturable microorganisms or, recently, as a result of the environmental DNA pool of nonculturable organisms, the metagenome; this increase was stimulated by an increasing demand for biocatalysts. Among these enzymes, coldadapted enzymes from psychrophiles, which are organisms with optimal temperatures for growth between 4 to 15°C, offer novel opportunities for biotechnological applications.

The aims of this study were:

1. Isolation and characterization of novel extremophiles, namely psychrophilic bacteria from seawater samples obtained from the area of Spitzbergen, in the Arctic.

2. Taxonomic classification of the newly isolated strain.

3. Screening for cold active esterases and lipases.

4. Cloning and characterization of novel cold-adapted and industrial relevant esterases and lipases.

2   MATERIALS AND METHODS

2.1   Isolation and characterization of aerobic bacteria.

2.1.1   Sample collection, media and culturing conditions

Arctic sea ice and seawater samples were collected in 1998 during an expedition in Spitzbergen, Norway. Samples were transported to the laboratory at an ambient temperature of 10 °C. 1 ml of each of the 50 liquid samples was incubated in a complex marine aerobic liquid medium.

The complex marine aerobic medium consisted of a basal medium supplemented with a solution of different carbon sources. The basal medium contained (g l^{-1}): NaCl 28.13 g; KCl 0.77 g; $CaCl_2$ x2H_2O 0.02 g; $MgSO_4$ x7H_2O 0.5 g; NH_4Cl 1.0 g; iron-ammonium-citrate 0.02 g; yeast extract 0.5 g; 10-fold concentration trace element solution (DSM 141) 1 ml; 10-fold concentration vitamin solution (DSM 141) 1 ml; KH_2PO_4 2.3 g; Na_2HPO_4 x2H_2O 2.9 g. The carbon source mixture solution contained (g l^{-1}): Na-acetate 0.5 g; Na_2-succinate 0.5 g; Na-pyruvate 0.5 g; DL-malate 0.5 g; D-mannitol 0.5 g and glucose 2 g. The final pH of the complex medium was 7.0. Incubation was carried out at 4 °C for about 4-7 days, before growth, and colonies on agar plates were observed. Colonies were selected on the basis of morphological differences. For the isolation of a pure culture, serial dilution and plating techniques were applied. The pure isolates were routinely cultivated on complex marine medium agar plates at 15 °C for 4 days.

2.1.2   Cellular characterization

Gram staining test was performed using the Hucker method (Gerhardt *et al*, 1994). For the sporulation test, cells were grown for up to 6 days in medium containing no carbon source but with 0.1 % (w/v) yeast extract. The presence of spores and the cell morphology was determined by phase-contrast microscopy (Zeiss/Axioskop).

2.1.3   Maximal temperature, pH and salt requirement for growth

Temperature growth range was tested between 4-35 °C and pH 7.0. The pH growth range was tested between pH 2-10 at 15 °C. The salt requirement was determined at different NaCl concentrations ranging from 0 to 10 % (w/v), with no change of the other salts concentrations at pH 7.0 and 15 °C.

Growth was measured by determining the optical density at 600 nm (1 - cm path length) using a Shimadzu UV spectrophotometer.

2.1.4   Substrate spectrum

Substrates utilization from the isolated aerobic strain was tested on api® 20 NE strips (20050) (Bio Merieux.Inc), api® 20 E strips (20 100/20 160) (Bio Merieux.Inc) and Biolog GN2 Microplate™. Api® 20 NE strips (20050) (Bio Merieux.Inc) and api® 20 E strips (20 100/20 160) were also used for testing the ability of the new strain to produce indol (tryptophane), acetoin and H_2S (from sodium thiosulfate), and the ability of the strain to reduce nitrate to nitrite and nitrite to nitrogen. Carbohydrate metabolism was also tested using api® 50 CH strips (300) (Bio Merieux.Inc).

The strips and micro-plates were inoculated with the cells suspension (cells were resuspended in NaCl 0.85 % (w/v) medium to a final OD_{600nm} 1.0) and incubated at 15 °C overnight.

37

A large number of carbon sources were tested: α cyclo-dextrin, dextrin, Tween 80, Tween 40, N-acetyl-D-glucosamine, α-D-glucose, maltose, sucrose, methyl pyruvate, D,L-lactic acid, succinic acid, bromo succinic acid, inosine, esculin ferric citrate, L-arabinose, potassium gluconate, malic acid, Trisodium citrate, glycogen, N-acetyl-D-galacto-samine, adonitol, capric acid, D-arabitol, cellobiose, I-erythol, d-fructose, L-fructose, D-galactose, gentiobiose, M-inositol, α-D-lactose, lactulose, D-mannitol, D-mannose, D-mellobiose, β-methyl D-glucose, D-psicose, D-raffinose, L-rhamnose, D-sorbitol, D-trehalose, turanose, xylitol, mono-methyl succinate, acetic acid, cis-aconitic acid, citric acid, formic acid, D-galactonic acid lactone, D-galacturonic acid, D-gluconic acid, D-glucosaminic acid, D-glucuronic acid, α-hydroxy-butyric acid, β-hydroxy-butyric acid, Y-hydroxy-butyric acid, p-hydroxy phenylacetic acid, itaconic acid, α-keto butyric acid, α-keto glutaric acid, α-keto valeric acid, malonic acid, propionic acid, quinic acid, D-saccharic acid, seabacic acid, succinamic acid, glucuronamide, alaninamide, D-alanine, L-alanine, L-alanyl-glycine, L-asparagine, L-aspartic acid, L-glutamic acid, glycyl-L-aspartic acid, glycyl-L-glutamic acid, L-histidine, hydroxy L-proline, L-leucine, L-ornithine, L-phenyl-alanine, L-proline, L-pyroglutamic acid, D-serine, L-serine, L-threonine, D,L-carnitine, Y-amino butyric acid, urocanic acid, uridine, thymidine, phenyl thylamine, putrescine, 2-amino ethanol, 2,3-butanediol, glycerol, D,L α-glycerol phosphate, glucose 1-phosphate, glucose 6-phosphate or phenylacetic acid.

2.1.5   *Enzyme production*

Amylase, arabinase, arabinoxylanase, protease, HE-cellulase, glucanase, dextranase, galactanase, galactomannanase, β-glucanase, pullulanase, curdlanase xylanase and xyloglucanase production by the isolated aerobic strain was tested on diffusion agar plates containing the base medium and 0.1 % (w/v) of one of the following substrates: red pullulan (pullulanase), azo-casein (protease) , AZCL–pullulan (pullulanase), AZCL–HE–cellulose (celulase), AZCL–arabinan (arabinase), AZCL–arabinoxylan (arabinoxylanase), AZCL–curdlan (curdlanase),

AZCL–amylose (amylose), AZCL–dextran (dextranase), AZCL–galactan (galactanase), AZCL–galactomannan (galactomannanase), AZCL–β-glucan (β– glucanase) AZCL–xylan (xylanase) and AZCL–xyloglucan (xyloglucanase). Substrate degradation was detected by clearing zone/color diffusion halo around the colonies after the isolated aerobic stain was grown on these substrates agar plates at 10 °C for 2-4 days. AZCL – polymers were purchased from Megazyme, Bray, Ireland.

2.1.6   *Fatty acids analysis*

Cells of the isolated strains were harvested from a two l culture sample by centrifugation and were used for fatty acids analysis. Fatty acids methyl ester (FAME) was performed according to the modified method of Lepage & Roy (Lepage & Roy, 1984). Total lipids were extracted according to Bligh & Dyer (Bligh & Dyer, 1959). The FAMEs were analyzed by capillary gas chrompack (Kohn *et al.*, 1996). A fused silica capillary column D23, 40 m (Fisons) was used for the separation of fatty acid species.

The chromatographic conditions were as follows: injector (PTV): 65 °C–270 °C split ratio 15:1; carrier gas: helium at a 40 cm s^{-1} flow. Column oven: initial temperature : 60 °C for 0.1 min; from 60 °C to 180 °C at 40 °C min^{-1}; 180 °C for 2 min; from 180 °C to 210 °C for 3 min; from 210 °C to 240 °C at 3 °C min^{-1}; 240 °C for 10 min. The spectra were recorded by a flame ionization detector at 280 °C.

2.1.7   *16S rDNA amplification and sequencing*

Cells of the isolated strains were harvested from a 500 μl culture sample by centrifugation and resuspended in 100 μl of water. A sample of 1 μl was used as a template for the amplification of the 16S rDNA. PCR-mediated amplification of the 16S rDNA was carried out according to Rainey & Stackebrandt (Rainey & Stackebrandt, 1993). PCR products were purified using the QIA quick PCR purification kit (Qiagen). Purified PCR products were directly sequenced using the Tag Dye Deoxy Terminator Cycle sequencing kit (Applied Biosystems). PCR product reactions were sequenced using Applied Biosystems model 373S DNA sequencer. Both strands of amplification products were sequenced using primers 8F, 518F and 1504R (Buchholz-Cleven *et al.*, 1997). The complete 16S rDNA sequence of the isolated strains was determined by the assembly of all sequence products using Vector NTI Advance 9.1 (Invitrogen) software.

2.1.8   *Phylogenetic analysis*

BLAST analysis was performed by NCBI online database of the 16S rDNA sequence to determine the phylogenetic grouping to which the isolated strains were most closely related. Reference sequences utilized in phylogenetic analysis were retrieved from NCBI database and aligned with the newly determined sequence of the isolated strains by using CLUSTAL W (1.83) software. The phylogenetic and molecular evolutionary analyses was performed with the neighbor joining method by using the software PHYLIP, version 3.65 (Felsenstien, 1995). The DNADIST program with Kimura-2 factor was used to compute the pairwise evolutionary distances for the above aligned sequences (Kimura, 1980). The topology of the phylogenetic trees was evaluated by performing a bootstrap (algorithm version 3.6 b) with 1000 bootstrapped trials. The phylogenetic trees were drawn using Tree View 32 software.

The 16S rDNA sequence data of the isolated aerobic strain was compared to available sequences of the microorganisms belonging to the genus *Pseudoalteromonas* and related microorganisms: *Pseudoalteromonas prydzensis* /U85855, *Pseudoalteromonas aliena* /AY387858, *Pseudoalteromonas tetraodonis* X82139, *Pseudoalteromonas carrageenovora* /X82136, *Pseudoalteromonas espejiana* /X82143, *Pseudoalteromonas atlantica* /X82134, *Pseudoalteromonas elyakovii* /AB000389, *Pseudoalteromonas distincta* /AF043742, *Pseudoalteromonas antarctica* /AF045560 /X98336, *Pseudoalteromonas gracilis* /AF038846, *Pseudoalteromonas nigrifaciens*

/X82146, *Pseudoalteromonas haloplanktis* /X67024, *Pseudoalteromonas undina* /X82140, *Pseudoalteromonas denitrificans* /X82138, *Pseudoalteromonas bacteriolytica* /D89929, *Pseudoalteromonas luteoviolacea* /X82136, *Pseudoalteromonas rubra* /X82147, *Pseudoalteromonas piscicida* /X82141 /X82215, *Pseudoalteromonas peptidolytica* /AF007286, *Pseudoalteromonas aurantia* /X82135, *Pseudoalteromonas citrea* /X82137, *Shewanella hanedai* /X82132, *Ferrimonas balearica* /X93021, *Vibrio cholerae* /X76337, *Aeromonas jandaei* /X74678, *Escherichia coli* /J01695, *Plesiomonas shigelloides* /X60418, *Moritella marina* /X82142, *Salinivibrio costicola* /X95527, *Alteromonas infernus* /X85175, *Alteromonas macleodii* /X85174 /X82145.

2.1.9 *G + C content of genomic DNA*

Cells of the isolated strains were harvested from a 2 l culture sample by centrifugation and were used for the determination of G + C content of the genomic DNA. The cells were disrupted with a French press and the DNA was purified by chromatography hydroxyapatit (Cashion *et al.*, 1977). The mol % G + C of genomic DNA was determined by high performance liquid chromatography (HPLC) (Shimadzu corp, Japan) (Mesbah & Whitman, 1989). The analytical column used was a VYDAC 201 SP 54, C18 5 μm (250 x 4.6 mm) equipped with a guard column 201 GD 54H (Vydac, Hesperia, USA).

2.1.10 *DNA – DNA hybridization*

DNA–DNA hybridization of the isolated aerobic strain with the strain *Pseudoalteromonas elyakovii* (LMG 14908) was carried out from 3 g cell mass of each strain. DNA was isolated using French press (Thermo spectroic) and was purified by chromatography on hydroxyapatite as described by Cashion et al (Cashion *et al.*, 1977). DNA–DNA hybridization was carried out as described by De Ley et al (De Ley *et al.*, 1970), with the modification described by Huss et al (Huss *et al.*, 1983), using a model 2600 spectrophotometer equipped with a model 2527-R thermoprogrammer and plotter (Gilford Instrument

Laboratories). Renaturation rates were computed with the TRANSFER. BAS program of Jahnke (Jahnke, 1992).

2.1.11 Bacterial strain, medium and growth condition

Pseudoalteromonas sp.. was grown aerobically in a complex marine medium consisting of basal medium supplemented with a solution of different carbon sources. The basal medium contained (gl^{-1}): NaCl 28.13 g; KCl 0.77 g; CaCl$_2$ x2H$_2$O 0.02 g; MgSO$_4$ x7H$_2$O 0.5 g; NH$_4$Cl 1.0 g; iron-ammonium-citrate 0.02 g; yeast extract 0.5 g; 10-fold concentration trace element solution (DSM 141) 1 ml; 10-fold concentration vitamin solution (DSM 141) 1 ml; KH$_2$PO$_4$ 2.3 g; Na$_2$HPO$_4$ x2H$_2$O 2.9 g. The carbon source mixture solution (g l^{-1}): Na-acetate 0.5 g; Na$_2$-succinate 0.5 g; Na-pyruvate 0.5 g; DL-malate 0.5 g; D-mannitol 0.5 g and glucose 2 g. The final pH of the complex medium was 7.0. Incubation was carried out at 15 °C for 2 days.

2.2   Construction and screening of a *Pseudoalteromonas* sp. nov. λ-phage gene library

2.2.1   Gene library construction

The lambda ZAP Express Predigested vector and ZAP Express Predigested Gigapack cloning kits (*Bam*H1/ CIAP-Treated) (Stratagene) were used for construction a *Pseudoalteromonas arctica* nov. λ-phage gene library. *E. coli* XL1-Blue MRF' and *E. coli* XLOLR were used as host cells as described by the supplier (Stratagene). Chromosomal DNA of *Pseudoalteromonas sp. nov.* was isolated by utilizing Qiagen`s genomic DNA isolation kit (Qiagen GmbH, Hilden). The genomic DNA was partially digested by *Sau*3A (New England, Biolabs, Inc.). Fragments were separated by agarose gel electrophoresis and 6–10 kb fragments were isolated by electroelution and phenol / chloroform extraction.

Pseudoalteromonas arctica nov. DNA fragments were ligated into the λ-ZAP express vector containing the phagemid pBK-CMV. After packaging of the ligation products the primary phage library was amplified in *E. coli* XL1-Blue MRF'. After amplification, phagemid pBK-CMV harboring the insert DNA was excised using helper phage ExAssist. The excised ExAssist population was transfected and stably established as plasmid in *E. coli* XLOLR cells.

2.2.2   *Screening for recombinant esterase*

The pBK-CMV plasmid, in *E. coli* XLOLR, contains IPTG–inducible T7 promoter and neomycin / kanamycin resistance ORFs. Therefore, positive clones harboring this plasmid were selected with kanamycin and IPTG. Esterase positive clones derived from *Pseudoalteromonas arctica* nov. gene bank were screened on LB / kanamycin (50 µg /ml) / IPTG (1 mM) agar plates. 1% Glyceryl tributyrate (Sigma) was added to select active recombinant clones of esterase. Incubation of screening plates was carried out at 20 °C for 3–4 days. The activity was observed when clearing zones (diffusion halos) around the active clones were seen as a result of tributyrin degradation (Fig.6).

Fig. 6 Clearing zones on LB agar plate observed around colonies expressing active esterase

2.2.3   *Plasmid isolation, sequencing and phylogenetic analysis*

Plasmids were isolated from the selected active clones using NucleoSpin plasmid isolation kit (Macherey–Nagel GmbH & Co. KG) and the insert size was determined by restriction analysis technique. DNA sequence was analyzed on an ABI automatic sequencer according to the Sanger method. Primer walking technique was used to determine the sequence of the insert containing the gene of interest. ORFs and genes were determined by analyzing the insert sequence with Vector NTI software (Invitrogen). The BLAST and FASTA algorithms were used to search the databases (Altschul *et al.*, 1990). Signal sequence prediction was carried out using SignalP 3.0 online software.

BLAST analysis was performed by NCBI online database to determine the phylogenetic group to which the new genes were most closely related. Reference amino acid sequences utilized in phylogenetic analysis were retrieved from NCBI database and aligned with the selected genes using CLUSTAL W (1.83) software.

2.2.4   *Esterase ORF/gene subcloning and expression*

PCR amplification was carried out using the Expand HiFi PCR kit (Roche Diagnostics, Germany) and temperature gradient thermoblock PCR system (Biometra, Germany) with the following temperature profile: 94 °C for 5 min and 30 cycles of 94 °C for 1 min, 59 ± 5 °C for 1min and 68 °C for 2 min, and then 68 °C for 15 min. The following primers were used in order to introduce an NdeI and an XhoI restriction sites at the beginning and end of the gene respectively.

Forward primer: (NdeI) CATATGCGACAAAAAGTATCTTTTAAAA

Reverse primer: (XhoI) CTCGAGTTCTACCAGTTCACTTACAATAACT

The subcloning of PCR–amplified fragments encoding esterase gene was carried out by using Acceptor vector kit (Novagen) where, the PCR–fragments were ligated into the expression vector pETBlue-1 (Fig. 7) and transformed into *E. coli* NovaBlue Singles (Novagen) competent cells.

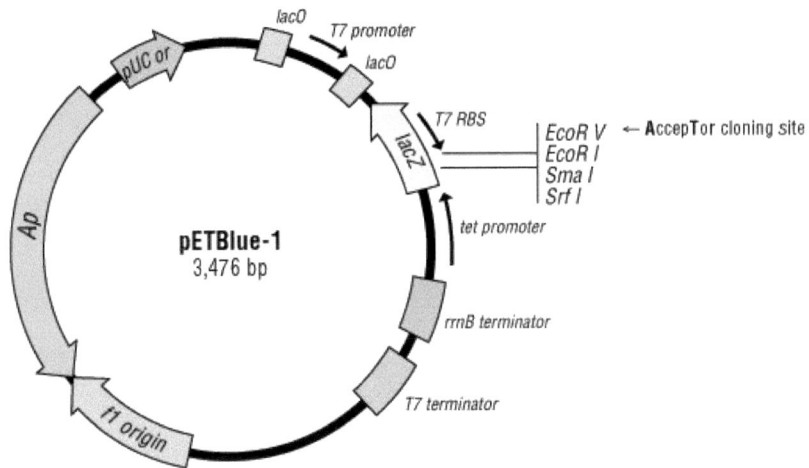

Fig. 7 pETBlue-1 Vector Map

The pETBlue™ vectors are designed to identify recombinants by traditional blue/white screening while also allowing T7lac promoter based expression of target genes. Screening is independent of expression because the T7lac expression promoter is in an opposed orientation relative to the *E.coli* promoter that mediates blue/white screening.

The transformed *E. coli* – cells obtained were screened for positive recombinants using blue white screening on LB agar plates supplemented with 50 µg/mL carbenicillin, 15 µg/mL tetracycline, 70 µg/mL X-gal, and 80 µM IPTG.

Plasmids were isolated from positive recombinant *E. coli* cells using NucleoSpin plasmid isolation kit (Macherey–Nagel GmbH & Co. KG) and were digested using *NdeI* and *XhoI* restriction enzymes overnight. pET24b + vector (Novagen) (Fig. 8) was digested using *NdeI* and *XhoI* restriction enzymes overnight. The digested plasmid and vector were both run on a 1% agarose gel and the obtained bands corresponding to the esterase gene and the vector were extracted using NucleoSpin Extract kit (Macherey–Nagel GmbH & Co. KG). The pET-24b vector carry an N-terminal T7•Tag® sequence plus an optional C-terminal His•Tag® sequence. The esterase gene was ligated in pET 24b vector in frame with the C-terminal His•Tag® sequence and transformed in *E.coli* - Tuner™(DE3 pLacI) Competent Cells (Novagen) for expression. The transformed cells were plated on LB agar plates containing 50 µg/mL kanamycin and 34 µg/mL chloramphenicol.

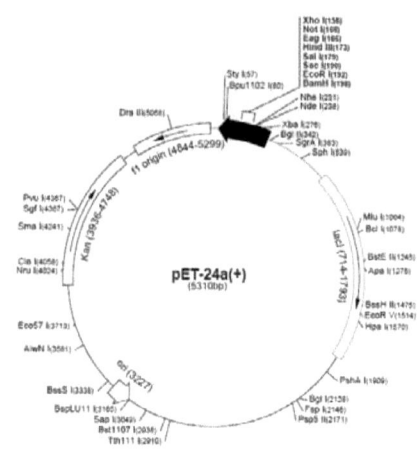

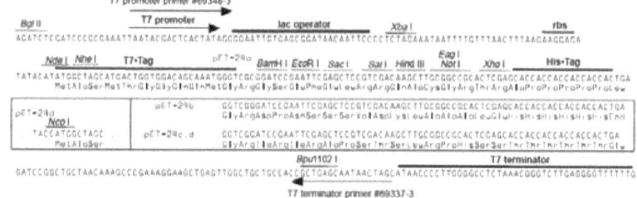

pET-24a-d(+) cloning/expression region

Fig. 8 pET 24 a-d (+) Vector Map

46

2.3   Enzyme assay

2.3.1   *Assay A: Spectrophotometric assay with p-nitrophenyl butyrate (p-NPB) as substrate*

Cleavage of *p*-NPB (Sigma, USA) was determined at 20°C in 25 mM Tris-HCl buffer pH 7.0, according to Winkler and Stuckmann (Winkler & Stuckmann, 1979).

Solutions for esterase assay

Solution A   37 mg *p*NPB (Sigma, USA) in 10 ml ethanol (10 mM stock)
Solution B   100 mg gum arabic in 90 ml 25 mM Tris-HCl buffer pH 7.0
Solution C   1 volume of solution A + 9 volumes of solution B

A buffered *p*-NPB emulsion (Solution C) was sonicated for 2 min at room temperature before the reaction was started by the addition of the enzyme. The reaction was stopped by cooling the mixture on ice for several minutes and then the addition of 100μl of 25% Na_2CO_3. The probes were centrifuged for 2 min at 13000 rpm and then optical density was measured at 410 nm. A blank measurement was taken immediately after enzyme addition.

1 unit of esterase activity is defined as the amount of enzyme, which releases 1 μmole *p*-nitrophenol per minute under test conditions. The activity toward other *p*-nitrophenyl esters was measured in the same manner, by using 1 mM of each substrate.

For the calculation of specific activities, protein concentrations in the extracts were determined using the Bradford method (Bradford, 1976).

Extinction coefficient $\sum 410$ for *p*-nitrophenol = $6.52 * 10^6$ cm^2 mole^{-1} at pH 7.0.

Calculation of the enzyme activity:

$$\text{Enzymes activity (U/ml)} = \frac{\Delta E / \Delta t}{d \cdot e} \cdot \frac{Vk}{Vp}$$

ΔE — change of absorption

Δt — change of time (min)

d — breadth of the cuvette (cm)

e — the molar extinction coefficient $(cm^2/mole)$

Vk — the general reaction volume (ml)

Vp — the volume of the enzyme (ml)

2.3.2   Assay B: Spectrophotometric assay with triacylglycerols and other substrates

The modified assay was described previously (Schmidt-Dannert *et al.*, 1994). The hydrolytic activity of the esterase was measured by a spectrophotometric assay using the formation of copper soaps for the detection of free fatty acids. Enzyme reaction was carried out under shaking for 12 hours at 20°C. The blank probe absorption, which was taken immediately after the start of the reaction, was measured at 430 nm. This method was used to determine the relative hydrolytic activity of the esterase towards different triacylglycerols and other substrates different from *p*-NP esters.

One unit (1 U) of esterase activity is the amount of enzyme needed to liberate 1 μmol of free fatty acid per minute under assay conditions.

2.4   Gel electrophoretic analysis

2.4.1   Native polyacrylamide electrophoresis

Native polyacrylamide Tris-Glycine gels containing a gradient of 4 to 20% polyacrylamide were from Novex (Novel Experimental Technology) (Invitrogen, The Netherlands). High- and low-molecular-weight markers (Amersham Biosciences, Germany) were used in order to determine the apparent molecular weight of the protein. Protein samples were mixed with standard sample buffer (Invitrogen, The Netherlands) and after 5 min applied onto the gel. The gels were run at 200 V for 12 h at room temperature.

2.4.2   *SDS polyacrylamide gel electrophoresis*

SDS-PAGE as described by Laemmli (Laemmli, 1970) was performed under reducing conditions with 10 % polyacrylamide gels by using a Bio-Rad mini-SUB CELL GT II electrophoresis unit (Bio-Rad, Germany). SDS high- and low-molecular-weight markers (Amersham Biosciences, Germany) were used in order to determine the apparent molecular weight of the proteins. Samples were mixed with loading buffer, without heat treatment, before loading. The gels were run at 100 Volts for 2 hours. The gels were then soaked in coomassie blue staining solution (0.15 % (w/v) coomassie brilliant blue R-250 in 45.5 % (v/v) methanol-8 % (v/v) acetic acid) for 30 min, this was followed by destaining for 3 hours in 25 % (v/v) ethanol-35 % (v/v) acetic acid.

2.5   Purification and characterization of recombinant esterase from E. coli Tuner™(DE3) cells

2.5.1   *Optimization of the expression of the recombinant esterase from E. coli Tuner™ (DE3) cells*

Optimization of expression conditions of the recombinant esterase from *E. coli* Tuner™(DE3) cells was carried out by culturing the cells in LB-kanamycin (50 μg/ml) liquid medium, in the presence and absence of IPTG (1 mM final concentration), at two incubation temperatures (30 and 37 °C), and incubation times (between 6 to 52 hours). The crude extract was obtained from the recombinant *E. coli* Tuner™ (DE3) cells and the esterolytic activity was tested as described in the enzyme assay (2.3.1). The extraction of the crude enzyme solution was carried out by harvesting the cells by centrifugation. The cells were resuspended in buffer (25 mM Tris-HCl containing 100 mM NaCl & 5 mM EDTA, pH 7.0) by keeping a constant ratio of 1:5 cell wet weight (g) to buffer

49

volume (ml). The biomass was disrupted by French Press (3 times at 2500 psi), and finally cell debris was removed by centrifugation (12.000 g, 20 min, 4 °C).

2.5.2 Purification of recombinant esterase from E. coli Tuner™ (DE3) cells

All purification steps were carried out at room temperature. Extraction was performed from 3 g *E.coli* Tuner™ (DE3) wet weight suspended in 15 ml 25 mM Tris HCl (pH 7.0). 1.5 ml Ni-NTA superflow Column (Qiagen) was used for purification. The column was equilibrated with 25 mM Tris-HCl (pH 7.0) buffer after which 1 ml sample from crude extract was loaded on the column. This was followed by a first wash with 25 mM Tris-HCl (pH 7.0) buffer and a second wash with 25 mM Tris HCl (pH 7.0) containing 25 mM imidazol. Elution was done with 25 mM Tris-HCl (pH 7.0) containing 250 mM imidazol. The eluted esterase (bearing a C-terminal His•Tag) was then dialyzed overnight at 4°C against 25 mM Tris-HCl (pH 7.0) buffer containing 100 mM NaCl and 5 mM EDTA. The purity of the obtained recombinant esterase was analyzed on a sodium dodecyl-sulfate (SDS) (10 %) polyacrylamide gel.

2.5.3 Influence of pH and temperature

Studies on the influence of the pH and temperature were carried out with the purified enzyme. The esterase activity was measured at a pH range from 4.0 to 13.0 using universal buffer (Britton & Robinson, 1931). All assays were performed at 20°C. To study the pH-stability of the enzyme, the esterase was incubated for 2 h at 20°C in universal buffer. To determine the influence of the temperature on the enzymatic activity, samples were incubated at temperatures from 0 to 40°C for 30 min (pH 7.5). Thermostability was investigated by incubation of the samples at temperatures from 4 to 60°C and pH 7.0. Samples were withdrawn at various time intervals (from 10 min to 24 h) and clarified by centrifugation. The enzymatic activity was measured as described in section (2.3.1).

2.5.4 *Effect of various compounds on esterase activity*

The effects of various substances on esterase activity was examined using Assay A after preincubation of the purified esterase with various reagents in different concentrations at 20°C for 120 min. For examination of the enzyme residual activity, 100 µl of preincubation mixture was mixed with 900 µl of substrate in 25 mM Tris-HCl buffer pH 7.0. The reaction was carried out for 20 min at 20°C, 1100 rpm. The enzyme activity without additional compounds was defined as 100%. Four groups of the various compounds (organic solvents, inhibitors, surfactants and metal ions) were studied for their effect on esterase activity.

Organic solvents:

Organic solvent	Concentration in preincubation mixture (v/v)	dissolved in
Tert butanol	50%, 99%	dH$_2$O
Ethanol	50%, 99%	dH$_2$O
Acetonitrile	50%, 99%	dH$_2$O
Isopropanol	50%, 99%	dH$_2$O
Pyridine	50%, 99%	dH$_2$O
Dimethylsulfoxide (DMSO)	50%, 99%	dH$_2$O
Acetone	50%, 99%	dH$_2$O
Dimethylformamide	50%, 99%	dH$_2$O
Methanol	50%, 99%	dH$_2$O
Hexadecane	99%	-
n-Hexane	99%	-
Heptane	99%	-
Isooctane	99%	-
Amylalcohol	99%	-
n-Decyl alcohol	99%	-
Toluol	99%	-
Benzene	99%	-

Inhibitors:

Inhibitor	Concentration in preincubation mixture	Inhibit	dissolved in
β-mercaptoethanol	10 mM	disulfide bond reducing reagent	dH$_2$O
Dithiotreitol (DTT)	10 mM	disulfide bond reducing reagent	dH$_2$O
p-chloro-mercuri-benzoate (PCMB)	10 mM	reagent masking SH-groups	dH$_2$O
Guanidine-hydrochloride	10 mM	denaturing reagent	dH$_2$O
Urea	10 mM	denaturing reagent	dH$_2$O
2-iodoacetate	10 mM	Cysteinproteases	dH$_2$O
Ethylenediamine tetraacetic acid (EDTA)	10% (w/v)	cation chelating reagent	dH$_2$O
Phenylmethylsulfonyl fluoride (PMSF)	10 mM	serine-modifying reagent	96 % (v/v) EtOH
Pefablock	10 mM	Serinproteases	dH$_2$O

Surfactants

Detergent	Concentration in preincubation mixture	resolved in
3-[(3-Cholamidopropyl) dimethylammonio]propanesulfonic acid (CHAPS)	10% (w/v)	dH$_2$O
Sodium dodecyl sulfate (SDS)	10% (w/v)	dH$_2$O
Triton X-100	10% (v/v)	dH$_2$O
Tween-80	10% (v/v)	dH$_2$O
Tween-20	10% (v/v)	dH$_2$O
Polyvinyl alcohol	10% (w/v)	dH$_2$O

Metal ions:

The salts containing the ions of Na^+, K^+, Ca^{2+}, Mg^{2+}, Cu^{2+}, Rb^{2+}, Ag^{3+}, Mn^{2+}, Co^{2+}, Sr^{2+}, Ni^{2+}, Al^{3+}, Fe^{3+}, Fe^{2+}, Zn^{2+}, Cr^{3+} were tested with a concentration of 5 mM in the incubation mixture.

2.5.5   *Substrate specificity*

The enzyme specificity was studied with p-nitrophenyl esters of varying alkyl chain lengths from C2 to C18 as described in assay A. The reactions were carried out at 20°C for 30 min.

The following substrates were tested as describes in assay A: 4-nitrophenyl benzoate, 4-nitrophenyl 2-(4-isobutylphenyl) propanoate, 4-nitrophenyl 2-phenylpropanoate, 4-nitrophenyl 3-phenylbutanoate, 4-nitrophenyl cyclohexanoate, 4-nitrophenyl 2-(3-benzoylphenyl) propanoate, 4-nitrophenyl-2-naphtoate, 4-nitrophenyl-1-naphtoate, 4-nitrophenyl-adamantanoate, 2-(4-isobutylphenyl)-N-(4-nitrophenyl) propanamide, (S)-4-nitrophenyl 2-(6-methoxynaphthalen-2-yl) propanoate (Table 6). The reactions were carried out at 20°C for 30 min.

The esterase activity was tested with the following compounds as described in assay B: tripalmitin (C16:0), tristearin (C18:0), triolein (C18:1), olive oil, Ethyl pelargonate, ethyl laurate, methyl caprylate, tolyl octanoate, ethyl caprylate, ethyl caproate, butyl laurate, vinyl acetate, vinyl butyrate and vinyl laurate. The reactions mixtures were incubated at 20°C for 12 h.

Table 6 Substrate specificity: The following substrates were tested as describes in assay A

#	Substrate	Chemical Structure
1	Ethyl nonanoate	
2	Ethyl Laurate	
3	Ethyl caprylate	
4	Ethyl caproate	
5	methyl octanoate	
6	n-Butyl laurate	
7	p- Tolyl Octanoate	$CH_3(CH_2)_5CH_2-\overset{O}{\overset{\|}{C}}-O-$ —CH_3
8	p-nitrophenyl acetate	
9	p-nitrophenyl butyrate	NO_2
10	p-nitrophenyl caproate	
11	p-nitrophenyl caprylate	
12	p-nitrophenyl palmitate	
13	p-nitrophenyl stearate	
14	4-nitrophenyl benzoate	

Table continued on next page

Table 6 Continued

#	Substrate	Chemical Structure
15	4-nitrophenyl 2-(4-isobutylphenyl) propanoate	Ibuprofen pnP ester
16	4-nitrophenyl 2-phenylpropanoate	
17	4-nitrophenyl 3-phenylbutanoate	
18	4-nitrophenyl cyclohexanoate	
19	4-nitrophenyl 2-(3-benzoylphenyl) propanoate	
20	4-nitrophenyl-2-naphtoate	
21	4-nitrophenyl-1-naphtoate	
22	4-nitrophenyl-adamantanoate	
23	2-(4-isobutylphenyl)-N-(4-nitrophenyl) propanamide	Ibuprofen amide
24	(S)-4-nitrophenyl 2-(6-methoxynaphthalen-2-yl) propanoate	Naproxen
25	Vinyl acetate	
26	Vinyl Propionate	
27	Vinyl butyrate	

2.5.6 Kinetic studies

Kinetic studies were performed with the recombinant enzyme (2 U/mg final concentration). In order to calculate the K_m and V_{max} the enzyme activity was tested as described in assay A but with various concentrations of pNPButyrate (0.05 - 2mM). The activity was tested at 20 °C, pH 7 and for 30 min.

3   RESULTS

3.1   *Pseudoalteromonas arctica* nov. aerobic, psychrotolerant bacterium isolated from Spitzbergen

3.1.1   *Physiological and morphological characteristics*

Enrichment cultures at pH 7 containing yeast extract, peptone and glucose inoculated with seawater sample from Spitzbergen showed bacterial growth after one week of aerobic incubation at 4 °C. Microscopy revealed the presence of short rod cells. After a number of serial dilutions, one culture was shown to exhibit the same characteristics and was selected as the culture for the strain *A* 37-1-2.

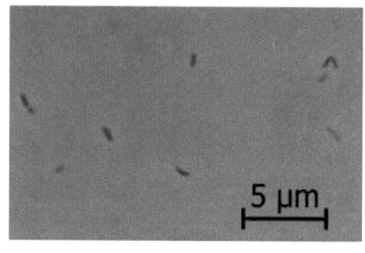

Cells of the strain *A* 37-1-2 are Gram negative, polar flagellated, short rods (0.15-0.2 μm wide and 0.4-0.75 μm long) and are motile. They occur as single cells, or in pairs. Spores were not detected. A comparison of the morphological characteristics of the strain *A* 37-1-2 (10) and the related strains (belonging to the *Pseudoalteromonas* genus): (1), *Pseudoalteromonas elyakovii* KMM 162 (Ivanova *et al.*, 1996); (2), *P. elyakovii* IAM 14595, a21, b11, b211, O22 (Sawabe *et al.*, 2000); (3), *P. espejiana* IAM 12640; (4), *P. citrea* KMM 216 (Ivanova *et al.*, 1998); (5), *P. carrageenovora* NCMB 302; (6), *P. atlantica* NCMB 301; (7), *P. distincta* KMM 638 (Ivanova *et al.*, 2000); (8), *P. haloplanktis* IAM 12915 and (9), *A. macleodii* IAM 12920 is shown in Table 7.

Table 7 Comparative characteristics of the strain A 37-1-2 and the strains of the Pseudoalteromonas genus. (1), Pseudoalteromonas elyakovii KMM 162; (2), P. elyakovii IAM 14595, a21, b11, b211, O22; (3), P. espejiana IAM 12640; (4), P. citrea KMM 216; (5), P. carrageenovora NCMB 302; (6), P. atlantica NCMB 301; (7), P. distincta KMM 638; (8), P. haloplanktis IAM 12915 and (9), A. macleodii IAM 12920 (Sawabe et al., 2000).

Character	Strain									
	1	2	3	4	5	6	7	8	9	A 37-1-2
Growth at:										
4°C	-	nd	+	-	+	-	-	+	-	+
37°C	-	nd	-	-	-	-	-	-	+	-
40°C	-	-	-	-	-	-	-	-	+	-
Production of:										
Amylase	+	+	+	+	-	+	-	-	+	-
Alginase	+	+	+	+	+	+	+	-	+	-
Agarase	-	-	-	+	-	+	-	-	-	-
Utilization of::										
D-Mannose	+	+	-	+	-	+	+	-	-	-
D-Galactose	+	+	+	+	-	+	-	-	+	+
D-Fructose	+	+	+	+	+	+	+	-	+	-
Sucrose, maltose	+	+	+	+	+	-	+	-	+	+
Melibiose	+	+	+	+	+	+	-	-	+	+
D-Lactose	+	+	+	+	+	+	-	-	+	-
D-Gluconate	-	+	-	-	-	-	-	+	+	+
N-Acetylglucosamine	-	nd	-	-	-	-	-	-	+	-
D-Mannitol	+	+	+	+	+	-	+	+	+	+
Succinate	+	+	+	+	+	-	+	+	+	-
Citrate	+	-	+	-	+	+	-	-	-	-
Meso-Erythritol	-	-	-	-	-	-	-	+	-	-
Glycerol	-	+	+	-	+	+	+	+	-	-
γ-Aminobutyrate	-	-	-	-	-	-	-	-	+	-
Xylose	+	+	+	+	-	-	-	-	+	-
Trehalose	-	-	+	+	-	+	-	+	-	-
Acetate	+	+	+	+	+	+	-	+	+	-
D-Glucosamine	-	-	-	-	-	-	-	+	-	-
Pyruvate	+	+	+	+	+	+	+	-	+	+
Alginate	+	+	+	+	+	+	+	-	+	-
Aconitate	+	-	-	-	-	-	-	-	-	-
D-Gluconate	+	-	-	-	-	-	-	-	-	+

+: positive, -: negative, nd: not determined.

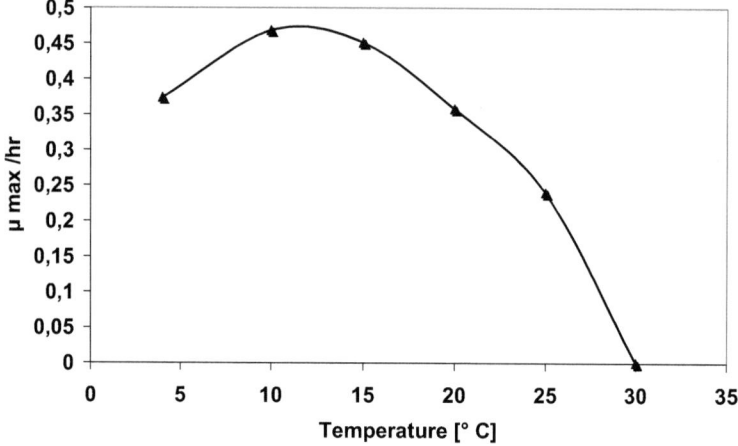

Fig. 9 Temperature range: The strain was cultivated in complex medium with 0.05% yeast extract and 0.2% glucose (pH 7.0) at a temperature range of 4-30°C. Maximal growth was observed at 10°C.

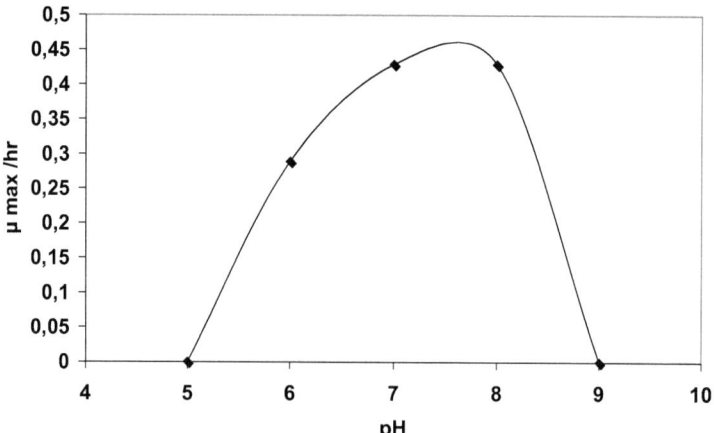

Fig. 10 pH range: The isolate was cultivated in complex medium with 0.05% yeast extract and 0.2% glucose (20°C) at a pH range of 5-9. Maximal growth was observed at pH 7-8.

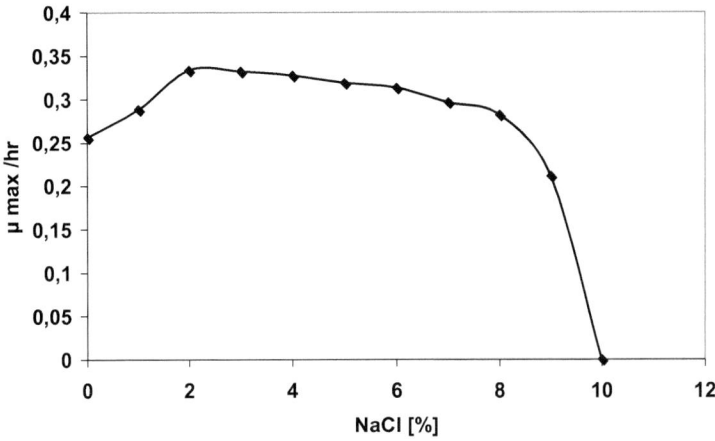

Fig. 11 Influence of NaCl on growth: The isolate was cultivated in complex medium with 0.05% yeast extract and 0.2% glucose (20°C, pH 7.0) at NaCl range between 0 and10%.

No growth was observed under anaerobic condition with glucose. Growth occurred only under aerobic conditions between 4 and 25°C. Optimal growth was observed at 10-15°C where the growth rate reached its maximum (Fig. 9). No growth was observed above 25°C. The strain A 37-1-2 required seawater and grew well at NaCl concentration of 0–9 % (w/v) with optimum at 2-3 % NaCl (w/v) (Fig. 11). The pH range for growth was 6–8, with an optimum at pH 7–8 (Fig. 10). Under the optimum conditions using 2 g^{-1} glucose, the growth rate was 0.47 h^{-1}.

3.1.2   Substrate spectrum

The strain A 37-1-2 grew on a variety of substrates. Growth was observed on: α-Cyclodextrin, Dextrin, D-Cellobiose, D-Galactose, α-D-Glucose, Maltose, D-Mannitol- D-Melibiose, Sucrose, Pyruvic acid methyl ester, D- Gluconic acid, L-Alanine, L-Alanylglycine, Glycyl-L-Glutamic acid, L-Serine, Esculine ferric citrate, and Capric acid.

No growth was observed on Glycogen, Tween80, Tween 40, N-Acetyl-D-Galactosamine, N-Acetyl-D-Glucosamine, Adonitol, L-Arabinose, D-Arabitol, i-Erythritol, D-Fructose, L-Fucose, Gentiobiose, m-Inositol, α-D-Lactose, Lactulose, D-Mannose, β-Methyl-D-Glucoside, D-Psicose, D-Raffinose, L-Rhamnose, D-Sorbitol, D-Trehalose, Turanose, Xylitol, Succinic acid Mono-Methyl-Ester, Acetic acid, Cis-Aconitic acid, Citric acid, Formic acid, D-Galacturonic acid, D-Glucosaminic acid, D-Glucuronic acid, α-Hydroxybutyric acid, β-Hydroxybutyric acid γ-Hydroxybutyric, p-Hydroxy phenylacetic acid, Itaconic acid, α-Keto butyric acid, α-Keto Glutaric acid, α-Keto Valeric acid, D,L-Lactic acid, Malonic acid, Propionic acid, Quinic acid, D-Saccharic acid, Sebacic acid, Succinic acid, Bromosuccinic acid, Succinamic acid, Glucuronamide, L-Alaninamide, D-Alanine, L-Asparagine, L-Aspartic acid, L-Glutamic acid, Glycyl-L-Aspartic acid, L-Histidine, hydroxy-L-Proline, L-Leucine, L-Ornithine, L-Phenylalanine, L-Proline, L-Phenylalanine, L-Proline, L-Pyroglutamic acid, D-Serine, L-Threonine, D,L-Carnitine, γ-Amino Butyric acid, Uronic acid, Inosine, Uridine, Thymidine, Phenylethylamine, Putrescine, 2-Aminoethanol, 2,3-Butanediol, Glycerol, D,L-α-Gylcerol phosphate, α-D-Glucose-1-phosphate, D-Glucose-6-phosphate, Malic acid, adipic acid, Trisodium citrate, Phenyl acetic acid, and Amygdalin.

The strain A 37-1-2 was not able to produce H_2S. Indole (Tryptophane) and acetoin were not produced. The strain was neither able to reduce nitrates to nitrites nor nitrites to nitrogen.

3.1.3   Enzymes production

The strain A 37-1-2 was able to produce several enzymes: pullulanase, protease, β-Glucanase, esterase, pectinase, β-glucosidase, and β-galactosidase. However, arabinase, amylase, xylanase, Alginase, chitinase, urease, and lipase were not detected.

3.1.4   Fatty acid analysis

The FAME composition of the strain A 37-1-2 (4) is shown in Table 8. The FAME profiles display only those fatty acids comprising ≥ 0.05 % of the total. Straight chain saturated FAME was 18.08 % total, terminally branched saturated FAME was 9.61 % total and monounsaturated FAME was 71.45% total. 16:0 straight chain saturated FAME (8.23 %) and 16:1 w7c monounsaturated FAME (58.60 %) were the most abundant FAME s found. The FAME composition of the strain A 37-1-2 and related strains (belonging to the genus *Pseudoalteromonas*) 1, *Pseudomonas haloplanktis* CECT 4188(=ATCC 14393); 2, *Pseudoalteromonas* sp. Strain CECT 579 (=ATCC 19262); 3, *Pseudoalteromonas atlantica* IAM 14164 (=ATCC 43555) are listed in Table 8.

Table 8 Comparative fatty acids composition (%) of the new isolated strain A 37-1-2 and related *Pseudoalteromonas* species. 1, *Pseudomonas haloplanktis* CECT 4188(=ATCC 14393); 2, *Pseudoalteromonas* sp. Strain CECT 579 (=ATCC 19262); 3, *Pseudolateromonas atlantica* IAM 14164 (=ATCC 43555).

Fatty Acid	1	2	3	A 37-1-2
Straight-chain fatty acids				
14:0	1.3		0.7	4.15
15:0	1.2	7.3	2.7	2.00
16:0	**20.1**	**11.6**	**19.4**	**8.23**
17:0	1.2	8.9	9.6	0.16
18:0			1.9	0.18
Terminally branched saturated fatty acids				
10:0 3OH		1.0	1.4	1.50
11:0 3OH		4.0	1.8	0.70
12:0 ISO 3OH	1.4			0.18
12:0 3OH	6.4	3.7	3.8	6.59
16:0 ISO	1.7			0.09
Monounsaturated fatty acids				
15:1 w8c		3.6		3.86
16:1 w7c	**34.8**	**19.5**	**22.0**	**58.60**
17:1 w8c	3.5	15.7	8.2	1.57

3.1.5   G+C content and DNA-DNA hybridization

The G + C content of the DNA from the strain A 37-1-2 was 39.0 mol%. The 16S rDNA sequence of the strain A 37-1-2 (1.499 kb) was analyzed and compared to available 16S rDNA sequences of organisms belonging to the genus *Pseudoalteromonas* (Fig. 12). The closest relationship was to *Pseudoalteromonas elyakovii* (identities: 1484/1489 - 99.66%), *Pseudoalteromonas citrea* (identities: 1469/1474 - 99.66%), *Pseudoalteromonas haloplanktis* (identities: 1484/1494 - 99.33%), *Pseudoalteromonas distincta* (identities: 1473/1481 - 99.45%), *Pseudoalteromonas agarovorans* (identities: 1475/1484 - 99.39%) and *Pseudoalteromonas gracilis* (identities: 1484/1494 - 99.33%) with an identity on 16S rDNA level of 99%. The DNA–DNA hybridization of the strain A 37-1-2 to the most related strain *Pseudoalteromonas elyakovii* showed only 47.6 % DNA–DNA similarity.

3.1.6   Description of Pseudoalteromonas arctica sp. nov.

Pseudoalteromonas arctica (arc' ti.ca. L. fem. adj. *arctica* from the Arctic, referring to the site were the type strain was isolated). Cells are straight, polar flagellated, rod shaped, gram negative, 0.15-0.2 µm wide and 0.4-0.7 µm long, occur singly, non spore forming and strictly aerobic. Colonies on agar medium are slightly orange, circular, smooth and convex. Temperature range for growth is 4–25 °C, with an optimum between 10 and 15 °C. No growth at 30 °C was detected. Range of pH for growth is 6–8, with an optimum at pH 7–8. Growth is observed at 0-9 % NaCl (w/v) concentration, with an optimum between 2 and 3 % (w/v). Growth is observed with α-Cyclodextrin, Dextrin, D-Cellobiose, D-Galactose, α-D-Glucose, Maltose, D-Mannitol- D-Melibiose, Sucrose, Pyruvic acid methyl ester, D- Gluconic acid, L-Alanine, L-Alanylglycine, Glycyl-L-Glutamic acid, L-Serine, Esculine ferric citrate, and Capric acid. The strain forms the following enzymes: pullulanase, protease, β-Glucanase, esterase, pectinases, β-glucosidase, β-galactosidase. The fatty acid methyl ester (FAME) composed of 18.08 % straight chain saturated FAME, 9.61 % terminally branched saturated FAME and 71.45

% monounsaturated FAME. Phylogenetic analysis reveals a close relationship to *Pseudoalteromonas elyakovii* with 99 % 16S rDNA composition identity and 47.6 % DNA-DNA similarity. The 16S rDNA (DQ787199) is 1.499 kb. The DNA base ratio is 39 mol % G + C. Habitat: artic sea water. The strain A 37-1-2 was isolated from seawater samples taken from Spitzbergen and deposited in the BCCM/LMG Bacteria Collection as *Pseudoalteromonas arctica* = LMG 23753.

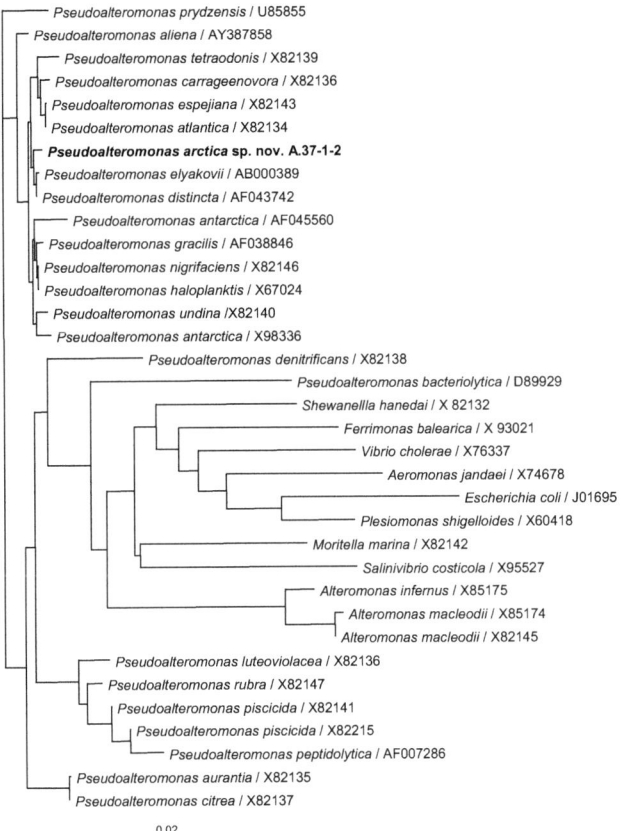

Fig. 12 Phylogenetic dendrogram based on 16S rDNA gene sequence comparison indicating the position of the new strain *Pseudoalteromonas* sp. nov. A 37-1-2 within the genus *Pseudoalteromonas*. performed by the neighbour-joining method using software from PHYLIP, version **3.65**; the DNADIST program with Kimura-2 factor was used to compute the pairwise evolutionary distances for the above aligned sequences , the topology of the phylogenetic tree was evaluated by performing a bootstrap (algorithm version 3.6 b) with 1000 bootstrapped trials. The tree was draw using Tree View 32 software. Bar correspond to 2 nucleotide substitutions per 100 nucleotides.

3.2 Cloning, sequencing, purification and characterization of an esterase from *Pseudoalteromonas arctica* sp. nov.

3.2.1 Cloning and sequencing of the 3.7-kb insert encoding an esterase from Pseudoalteromonas arctica sp. nov.

Ten thousand phagemid clones obtained from a λZAP (Stratagene) genome bank were screened for esterase activity at 30 °C for 3-4 days as described in Materials and Methods section 2.2.1. Several clones showed clearing zones on tributyrin plates. The insert sizes of the clones were about 3 to 4 kb. The 4 kb clone was sequenced using primers walking technique. Two forward and reverse walking with 4 constructed primers (Table 9) were necessary to determine the complete sequence of the insert. M13 forward and M13 reverse standard primers were used to start the primer walking. Complete sequence revealed two open reading frames (ORF): *Est 37* and *ORF2* as shown in (Fig. 13). The ORF *Est 37* (1.203 kb) encodes a protein Est 37 of 400 amino acids with a predicted molecular mass of 44 kDa. An Interpro scan by EBI database showed that this ORF/gene has two domains; an esterase and an OsmC (Osmotic shock response protein) like domain (Fig. 14).

Table 9 Oligonucleotides used for the primer walking of the 4-kb insert encoding esterase gene. M13 forward and M13 reverse standard primers were used to start the primer walking.

Walk	Primer direction	Primer position	Sequences
	F = forward R = reverse	(bp)	(5´- 3´)
1	F	3108-3127	GTGATAAACCTAAACCTTCA
	R	616-635	ATCTGCAAGCCAAGTATGAT
2	F	2509-2529	GGATAATGTAGGATACGTAAC
	R	1212-1236	CCGCAGATACTAAATCTTGAATGTT

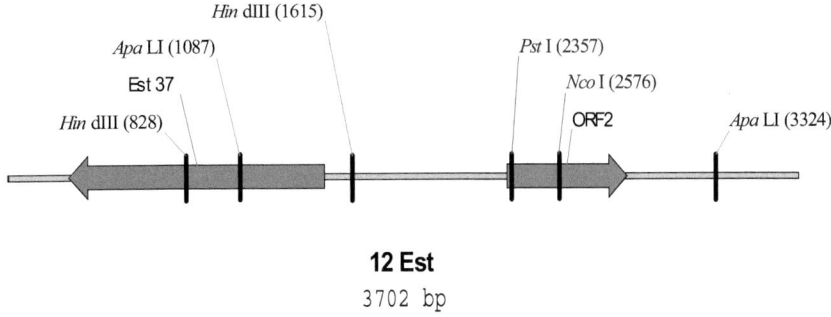

12 Est

3702 bp

Fig. 13   Map of the 3.7-kb insert encoding the esterase from *Pseudoalteromonas arctica*. The whole insert (3.702 kb) revealed a G+C content of 46.82 % and one open reading frame (ORF) encoding the esterase gene (marked in orange). The gene, 277-bp to 1476-bp, has a size of 1203 bp, encodes a protein of 400 amino acids with a predicted molecular mass of 45 kDa

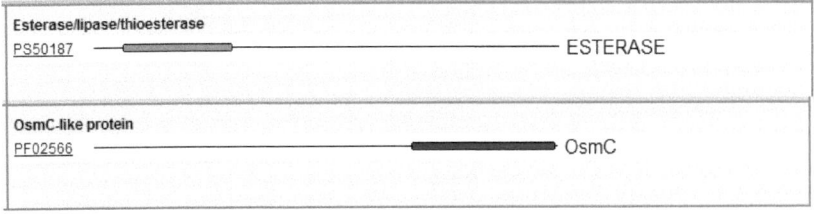

Fig. 14 EBI interpro scan detected two putative conserved domains, an esterase and an OsmC (stress protein) like domain.

3.2.2   Esterase gene analysis

The esterase gene *Est 37* (Fig 15) from *Pseudoalteromonas arctica* sp. nov. has an ATG (M) start codon, a TAA stop codon, is 1.203 kb long, and encodes an enzyme (Est 37) (400 aa) with a predicted molecular mass of 44 kDa and a predicted pI of 6.23. No signal sequence was detected (SignalP 3.0 Server). A hypothetical Shine-Dalgarno-like sequence with a weak motif (AAGG) (Ma *et al.*, 2002) was found at 7 to 10 bp upstream from the predicted translational start codon.

The esterase protein sequence exhibits 90 % amino acid identity with a hypothetical protein from *Pseudoalteromonas haloplanktis* containing similar esterase and OsmC domains, 55% identity to an OsmC like protein from *Oceanospirillum* sp. MED92, and 62% identity to an α/β hydrolase from *Legionella pneumophila*.

The highly conserved five-residue motif typical of all α/β hydrolases (GxSxG) was detected at position 104 to 108. A catalytic triad consisting of one serine, one histidine and one aspartic acid residue was detected at positions 106, 225 and 196 respectively. It is therefore proposed that Est 37 belongs to the serine hydrolase family with a potential catalytic triad comprising Ser_{106}, found within the GHSLG consensus sequence, known as the substrate-binding site, Asp_{196} and His_{225} (Fig. 16).

```
    ATAAGGCTCCCC
      M   R   Q   K   V   S   F   K   S   G   D   L   E   L   A   G   Q  ·
   1 ATGCGACAAA AAGTATCTTT TAAAAGCGGC GATTTAGAAC TTGCCGGCCA
      ·  L   E   L   P   S   G   D   V   K   F   Y   A   L   F   A   H   C  ·
  51 ACTTGAACTT CCCTCTGGTG ACGTTAAGTT TTACGCGCTA TTTGCACACT
      ·  F   T   C   G   K   D   I   A   A   A   T   R   I   S   R   A
 101 GCTTTACCTG CGGTAAAGAC ATTGCAGCAG CCACTCGTAT TAGCCGAGCT
      L   T   Q   Q   G   I   A   V   L   R   F   D   F   T   G   L   G  ·
 151 TTAACACAAC AAGGCATTGC CGTACTACGT TTCGACTTTA CCGGTTTAGG
      ·  N   S   D   G   D   F   A   N   S   N   F   S   S   N   I   Q   D  ·
 201 TAATAGCGAT GGCGACTTTG CTAACAGTAA CTTTTCATCA AACATTCAAG
      ·  L   V   S   A   A   N   H   L   R   E   H   F   A   A   P   Q
 251 ATTTAGTATC TGCGGCAAAT CATTTACGTG AGCATTTTGC GGCGCCGCAA
      L   L   I   G   H   S   L   G   G   A   A   V   L   A   A   A   E  ·
 301 CTACTCATTG GCCACAGTTT AGGCGGGGCT GCCGTTCTTG CTGCTGCGGA
      ·  H   I   L   E   V   S   A   I   T   T   I   G   A   P   S   D   A  ·
 351 GCATATTCTT GAAGTATCGG CTATTACAAC CATTGGTGCA CCGTCAGATG
      ·  Q   H   V   A   H   N   F   E   A   H   L   D   E   I   N   A
 401 CGCAGCACGT AGCGCATAAT TTTGAAGCAC ACCTTGATGA AATTAACGCA
      A   G   E   A   K   V   N   L   A   G   R   E   F   T   I   K   K  ·
 451 GCAGGTGAAG CTAAAGTAAA CTTAGCCGGC CGTGAATTTA CCATTAAAAA
      ·  Q   F   I   D   D   I   A   K   Y   D   K   S   H   I   S   K   L  ·
 501 GCAATTTATT GACGATATAG CCAAGTACGA TAAAAGCCAC ATAAGTAAAC
      ·  K   R   A   L   L   V   M   H   S   P   I   D   A   T   V   N
 551 TTAAGCGCGC ATTATTAGTA ATGCACTCCC CTATTGATGC GACGGTAAAT
      I   S   E   A   E   K   I   Y   A   S   A   K   H   P   K   S   F  ·
 601 ATTAGTGAAG CTGAAAAAAT TTATGCATCA GCCAAGCATC CTAAAAGCTT
```

```
          · I   S   L   D   N   A   D   H   L   L   T   N   K   N   D   A   D ·
      651 TATTAGCCTA GATAACGCCG ATCACCTTTT AACAAATAAA AACGATGCCG
          · Y   A   A   Q   I   I   A   T   W   A   N   R   Y   V   K   Y
      701 ACTACGCAGC ACAAATAATT GCAACGTGGG CAAACCGTTA TGTTAAGTAC
            D   K   T   K   Y   T   A   S   L   T   G   G   N   V   L   V   E ·
      751 GACAAAACTA AATACACGGC AAGTTTAACG GGTGGCAATG TACTCGTTGA
          · E   K   D   H   V   F   T   Q   H   V   S   T   K   D   H   T   W ·
      801 AGAAAAAGAC CATGTATTTA CTCAGCACGT AAGTACAAAA GATCATACTT
          · L   A   D   E   P   I   K   V   G   G   K   N   L   G   P   D
      851 GGCTTGCAGA TGAGCCAATA AAAGTAGGTG GTAAAAACTT AGGTCCTGAT
            P   Y   H   H   L   L   A   G   L   G   A   C   T   A   M   T   L ·
      901 CCGTATCATC ACTTATTAGC GGGGCTTGGT GCCTGTACGG CCATGACACT
          · R   M   Y   A   T   R   K   N   L   P   L   E   H   V   K   V   E ·
      951 GCGTATGTAC GCTACACGTA AAAACTTACC ACTGGAGCAT GTAAAAGTAG
          · L   D   H   T   R   D   Y   N   K   D   C   D   D   C   E   Q
     1001 AGCTTGATCA CACTCGCGAT TACAACAAAG ATTGCGACGA TTGTGAGCAA
            T   G   N   L   E   A   I   T   R   K   I   T   L   R   G   D   L ·
     1051 ACAGGTAACC TTGAAGCAAT TACCCGTAAA ATCACCTTAC GTGGCGACTT
          · T   E   P   Q   R   Q   R   L   L   E   I   A   D   K   C   P   V ·
     1101 AACAGAGCCA CAACGCCAGC GTTTACTCGA AATAGCCGAC AAATGCCCTG
          · H   K   T   L   H   N   N   P   V   I   V   S   E   L   V   E
     1151 TGCATAAAAC ACTACATAAT AACCCAGTTA TTGTAAGTGA ACTGGTAGAA
            *
     1201 TAACTG
```

Fig. 15 Nucleotide sequence and deduced amino acid sequence of the esterase from *Pseudoalteromonas arctica*. The start codon is (ATG) and the stop codon (TAA) is marked by an asterisk. The hypothetical Shine-Dalgarno-like sequence is underlined. The highly conserved GxSxG motif typical of α/β hydrolases is given in bold underlined letters. The catalytic triad consisting of one serine, one histidine and one aspartic acid residue detected at positions 106, 225 and 196 respectively is in bold italics.

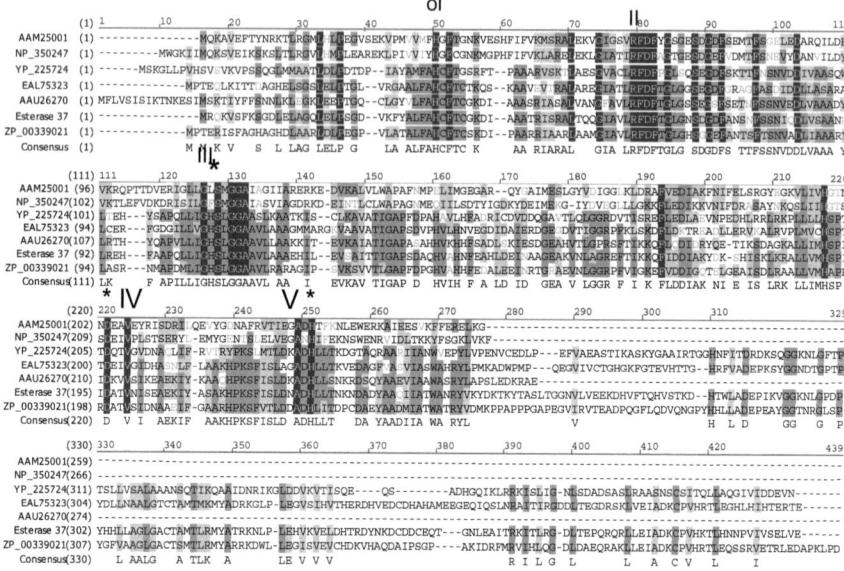

Fig. 16 Comparison of microbial α/β hydrolase sequences. The alignment of the deduced amino acid sequence of *Pseudoalteromonas arctica* Est 37 and homologous α/β hydrolases. The accession numbers of the aligned sequences are for the following organisms: AAM25001, α/β hydrolase from *Thermoanaerobacter tencongensis* MB4; NP_350247, α/β hydrolase from *Clostridium acetobutylicum* ATCC 824; YP_225724, α/β hydrolase from *Corynebacterium glutamicum* ATCC 13032; EAL75323, α/β hydrolase from *Erythrobacter litoralis* HTCC 2594; AAU26270, α/β hydrolase from *Legionella pneumophila* subsp. *pneumophila* str. Philadephia 1; ZP_0033902, α/β hydrolase from *Silicibacter* sp. TM 1040. The accession numbers are indicated to the left of the amino acid sequences. Identical residues have a black background and similar residues have a grey background. o marks oxyanionhole, * marks the catalytic triad, I and II, III, IV and V mark highly conserved blocks.

The comparison of the amino acid sequences of the *Pseudoalteromonas arctica nov.* esterase with the reported esterases from various strains (Table 10) demonstrated, that this esterase does not have high similarity to another known published esterase sequences. The identity % between the known characterized esterases and Est 37 was summarized in the Table 10. Est 37 sequence has less than 10% identity to the amino acid sequences of known esterases. It does not belong to any of the known α/β hydrolase families as well.

Table 10 Comparison of sequences of esterase from different microorganisms. Sequence pair distances of the esterases, using Clustal W 1.83 method with PAM250 residue weight table. The protein sequences are from: 1. *Streptomyces scabiei* 2. *Moraxella* sp. 3. *Pseudomonas aeruginosa* 4.*Pseudomonas* sp. B11-1 5. *Salmonella typhimurium* 6. *Photorhabdus luminescens* 7. *Enterobacter cloacae* 8. *Delftia acidovorans* 9. *Bacillus* sp. BP-23 10. *Bacillus* sp. BP-7 11. *Bacillus subtilis* 12. *Bacillus licheniformis* 13. *Archaeoglobus fulgidus DSM* 4304. 14. *Psychrobacter* sp. Ant300 15. *Moraxella* sp. 16. *Acinetobacter* sp. No.6 17. *Pseudomonas fluorescens* 18. *Sulfolobus acidocaldarius* 19. *Escherichia coli* K12 20. *Acetobacter pasteurianus* 21. *Moraxella* sp. 22. *Psychrobacter immobilis*. 23.*Pseudoalteromonas arctica* nov.

	1.AAA26743	2.CAA37220	3.AAB61674	4.BAC21259	5.AAC38796	6.CAA47020	7.AAL82802	8.BAA76305	9.CAB42083	10.CAB93516	11.P37967	12.CAC51386	13.AAB89533	14.BAD06009	15.CAA37862	16.BAB68337	17.AAC60471	18.AAC67392	19.AAC76437	20.BAA25795	21.CAA37863	22.CAA47949	23.Est 37
1. AAA26743		11	9	9	9	9	13	9	10	12	11	8	6	8	8	7	8	6	7	8	9	8	8
2. CAA37220			7	6	9	7	7	9	7	8	9	8	4	7	8	6	6	6	6	12	9	8	6
3. AAB61674				65	22	21	8	10	12	12	13	11	7	9	8	8	8	7	6	9	9	8	5
4. BAC21259					19	18	8	9	11	10	10	9	6	9	8	7	6	8	7	9	8	8	6
5. AAC38796						41	5	7	7	6	6	4	3	5	6	8	5	7	6	7	9	8	7
6. CAA47020							6	7	6	5	5	5	5	6	6	6	6	6	6	5	7	8	6
7. AAL82802								31	29	30	31	26	7	7	8	10	8	10	6	8	9	7	8
8. BAA76305									27	30	29	25	8	11	11	9	8	8	6	8	8	6	10
9. CAB42083										48	49	44	8	11	10	8	8	8	5	10	8	9	7
10.CAB93516											83	60	10	11	10	6	6	6	7	7	9	9	7
11.P37967												58	9	11	10	8	7	7	6	7	11	9	7
12.CAC51386													7	8	9	7	8	6	6	7	9	7	6
13.AAB89533														29	30	14	9	8	6	9	4	4	4
14.BAD06009															53	13	6	8	6	7	9	8	6
15.CAA37862																14	8	6	6	9	8	8	6
16.BAB68337																	8	8	7	8	8	8	8
17.AAC60471																		7	7	9	8	7	8
18.AAC67392																			22	14	12	12	9
19.AAC76437																				17	11	13	7
20.BAA25795																					14	16	9
21.CAA37863																						86	6
22.CAA47949																							6
23.Est 37																							

The amino acid sequence of Est 37 was analyzed using Vector NTI software and the results are summarized in the Tables 11 and 12. According to analysis of the amino acid composition, there are 37% hydrophobic residues and 24% polar residues. Among the charged residues, 13% are acidic and 11% are basic.

3.2.3 Subcloning of the Est 37 gene

The ORF *Est 37* (Fig. 13) was subcloned into expression vector pET-24 b+ and expressed in *E. coli* Tuner (DE3 pLacI) cells as described in section 2.2.4. The forward and reverse primers were used for the amplification of Est 37 gene. Esterase activity could be detected as clearing zones around the active clones after 2-3 days of growth on LB/chloramphenicol/kanamycine agar plates containing 1 % (w/v) tributyrin at 30 °C.

Table 11 Analysis of the amino acid sequence of Est 37

Analysis	Entire Protein
Length	400 aa
Molecular Weight	43959.01 m.w.
1 microgram =	22.748 pMoles
Molar Extinction coefficient	24900
1 A[280] corr. to	1.77 mg/ml
A[280] of 1 mg/ml	0.57 AU
Isoelectric Point	6.23
Charge at pH 7	-6.41

Table 12 Amino acid composition of Est 37

Amino Acid(s)	Number count	% by weight	% by frequency
Charged (RKHYCDE)	131	38.32	32.75
Acidic (DE)	51	13.87	12.75
Basic (KR)	43	13.11	10.75
Polar (NCQSTY)	96	24.18	24.00
Hydrophobic (AILFWV)	150	35.22	37.50
A Ala	47	8.19	11.75
C Cys	6	1.42	1.50
D Asp	29	7.55	7.25
E Glu	22	6.33	5.50
F Phe	14	4.52	3.50
G Gly	23	3.38	5.75
H His	21	6.37	5.25
I Ile	24	6.16	6.00
K Lys	28	8.00	7.00
L Leu	41	10.52	10.25
M Met	4	1.17	1.00
N Asn	20	5.17	5.00
P Pro	12	2.70	3.00
Q Gln	13	3.71	3.25
R Arg	15	5.11	3.75
S Ser	22	4.52	5.50
T Thr	25	5.82	6.25
V Val	22	5.04	5.50
W Trp	2	0.80	0.50
Y Tyr	10	3.54	2.50
B Asx	49	12.71	12.25
Z Glx	35	10.04	8.75
X Xxx	0	0.00	0.00

3.2.4   *Purification of the recombinant esterase after expression in E. coli tuner DE3 placI*

Expression of the recombinant esterase in *E. coli* Tuner DE3 placI was studied with and without induction (2 mM IPTG) (Fig. 17), at 30 °C and 37 °C and at different incubation times (6–55 hours). Optimal expression conditions for the recombinant enzyme production were determined by measuring the enzyme activity (U/mg) of the crude extract from the *E.coli* Tuner DE3 placI clone (as described section 2.3.1). The highest specific activity (0.5 U/mg) was reached after 24 hour growth on LB-kanamycin medium, with induction with 2 mM IPTG at 30 °C.

Purification of recombinant esterase (bearing a C-terminal His•Tag) was carried out from the crude extract of *E.coli* Tuner DE3 cells grown at optimal expression conditions as described in section 2.5.2. The results of the purification procedure are summarized in Table 13. After Ni-NTA superflow (Qiagen) Column chromatography the recombinant esterase was purified 3.3 fold, with a specific activity of 1.7 U/mg, and yield of 96 %. Proteins from the purification steps were separated using SDS-PAGE-10 % (Fig. 18).

Table 13 Purification of the recombinant esterase of *Pseudoalteromonas arctica* after expression in *E. coli* Tuner DE3 placI.

Purification Step	Fraction Volume [ml]	Total Protein [mg]	Total Activity [U]	Specific Activity [U/mg]	Yield [%]	Purification factor [fold]
Cell Extract	1	18.3	9.5	0.5	100	1
HisTag Column	6	5.3	9.2	1.7	96	3.3

* After the aerobic growth of E. coli Tuner DE3 placI at 30 °C in 1 L LB medium, culture cells were centrifuged (3 g of wet weight cells), resuspended with 15 ml Tris-NaCl buffer (pH 7.5), disrupted with French press and the supernatant was used for purification. One unit of esterase is defined as the amount of enzyme that releases 1 μmol of p-nitrophenol per min under assay conditions.

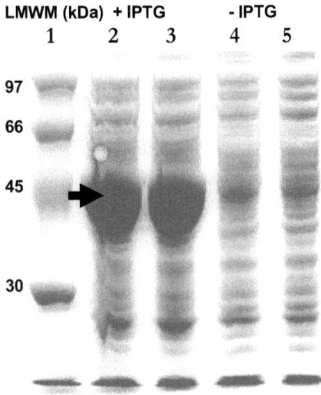

Fig. 17 Gel electrophoretic analysis (10% SDS – PAGE) of the *E. coli* Tuner DE3 placl crude extract with and without 2mM IPTG induction. Upon induction with IPTG the expression of the recombinant esterase can be clearly seen as a thick protein band (→) with an apparent molecular mass of 45 kDa. Lane 1: Low molecular weight marker, lanes 2&3: cell extract after induction with 2mM IPTG, lanes 4&5: cell extract without IPTG induction.

The samples after Ni-NTA superflow (Qiagen) Column chromatography revealed one protein band with apparent molecular mass of 45 kDa (Fig. 18).

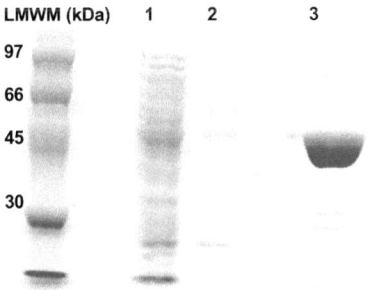

Fig. 18 Gel electrophoretic analysis (10 %-SDS-PAGE) of the purified recombinant esterase. Purification of the recombinant esterase. Lane 1, First wash, 25 mM Tris buffer; lane 2, 2nd wash - 25 mM Tris buffer + 25 mM imidazol; lane 3, Eluted protein - 25 mM Tris buffer + 250 mM imidazol.

3.2.5   Biochemical properties of recombinant esterase

Recombinant esterase was active between 0 °C and 30 °C. The temperature optimum was 25 °C and a rapid decrease in esterase activity was observed above 30 °C. 50% activity was retained at 0°C (Fig. 19-A). Recombinant esterase showed activity over a pH range of 6.5-9.0, with an optimum at pH 7.5-8.0 (Fig. 19-B). The recombinant esterase has an activation energy E_a of 5.1 kcal/mol (from 0 to 25°C) (Fig. 20). The influence of temperature on the stability of the recombinant esterase was examined by measuring the enzymatic activity after incubation at temperatures ranging from 4 up to 60°C (Fig. 21). The enzyme is stable at temperatures of 4 °C and 10 °C. It lost around 35% of activity after 24 hours of incubation at 20°C. It has a half life of 15 hours at 30 °C and 5 hours at 40°C. The influence of pH on the stability of the recombinant esterase was examined by measuring the enzymatic activity after incubation at pHs ranging from 5 up to 12 (Fig. 22). The enzyme is stable at alkaline rather than at acidic pHs.

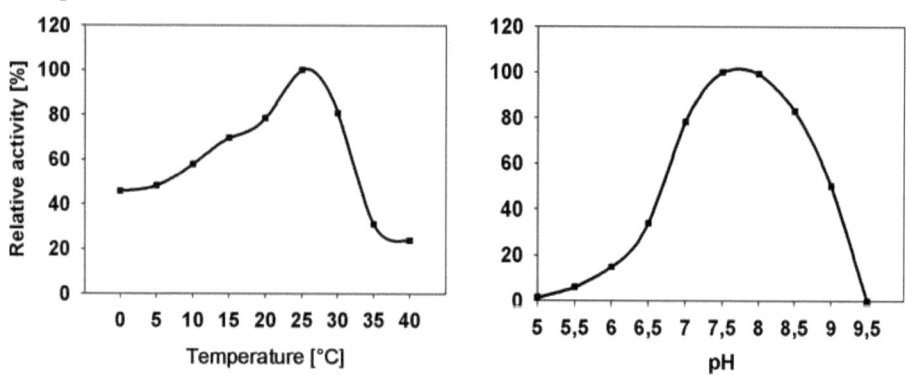

Fig. 19 Influence of temperature (A) and pH (B) on the activity of the recombinant esterase from *Pseudoalteromonas arctica*. For the determination of optimum temperature the recombinant (2 U/mg) enzyme was incubated in Tris-HCl buffer (pH 7.5) for 30 min at different temperatures. For the determination of pH optimum recombinant enzyme was incubated (25 °C) in universal buffer (pH 5 to 9.5) for 30 min at different pH values. PNP butyrate was used as substrate and the enzymatic reaction was carried out as described in section 2.3.1.

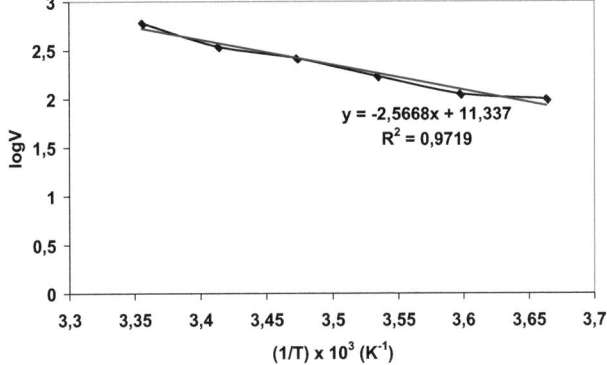

Fig. 20 Arrhenius plot for the determination of the recombinant esterase's activation energy (E_a). Logarithms of Vmax values were plotted as a function of the reciprocal of the absolute temperature. Linear regression gives a straight line with a slope of - E_a/R = -2.5668. (R is the ideal gas constant in joules per mole Kelvin). E_a = 21.3 kJ/mole = 5.1 kcal/mol.

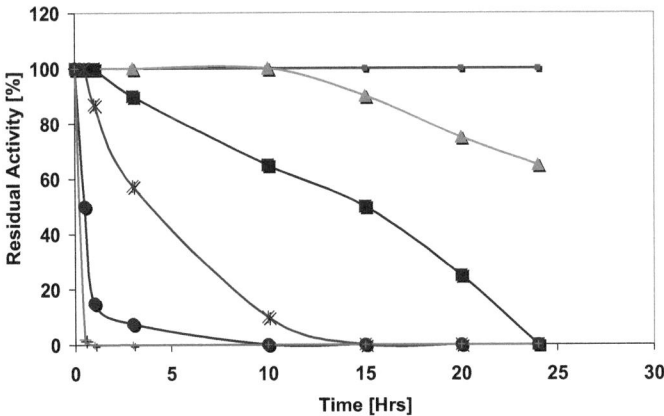

Fig. 21 Thermostability of the recombinant esterase from *Pseudoalteromonas arctica*. The enzyme (2 U/mg) was incubated at 4 °C (⋅), 10 °C (•), 20 °C (▲), 30 °C (■), 40 °C (x), 50°C (●)and 60 °C (+) in 25 mM Tris buffer pH 7.5. Samples were taken and tested for esterase activity. PNP butyrate was used as substrate and the enzymatic reaction was carried out as described in section 2.3.1.

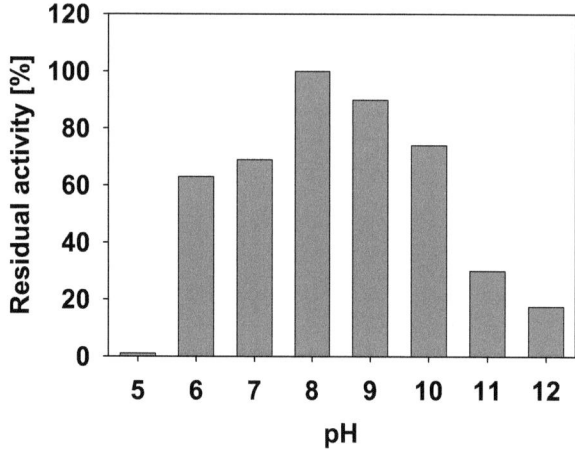

Fig. 22 pH stability of the recombinant esterase from *Pseudoalteromonas arctica*. The enzyme (2 U/mg) was incubated in 50 µl universal buffer for 2 hours at a range of pH of 5 to 12. Samples were taken and tested for esterase activity as described in section 2.3.1. PNP butyrate was used as substrate.

3.2.6   Effect of metal ions and other reagents on recombinant esterase activity

The effect of metal ions, surfactants, inhibitors and organic solvents on *Pseudoalteromonas arctica* sp. nov. recombinant esterase is shown in Fig. 23, 24, 25, and 26 . Divalent and monovalent cations affected the activity of the recombinant enzyme as follows: the recombinant esterase was inhibited by 10 mM of Al^{+2} (100% Inhibition), Cu^{+2} (100% Inhibition), Fe^{+2} (100% Inhibition), Ni^{+2}, Cr^{+2}, and Co^{+2}; whereas, Ca^{+2}, Mg^{+2}, Se^{+2}, Mn^{+2} and K^{+1} had no or limited effect. Na^{+1} increased the activity by 30% (Fig. 23).

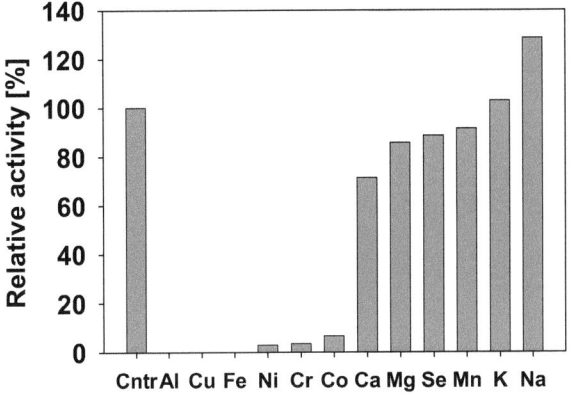

Fig. 23 Effect of various metal ions on lipase activity was examined using Assay A (s. 2.3.1). The esterase was incubated with various metal ions (10 mM) for 120 min at 20°C. For examination of the enzyme residual activity, 100 µl of preincubation mixture were mixed with 900 µl of substrate (PNP butyrate) in 25 mM Tris-HCl buffer pH 7.0. The reaction was carried out for 20 min at 20°C, 1100 rpm.

The effect of various inhibitors was also investigated (Fig 24, 25). At concentrations of 10 mM, urea and guanidine hydrochloride showed no or mild inhibitory effect on the activity of the recombinant esterase. Disulfide bond reducing reagents such as β-Mercaptoethanol reduced the activity down to 30% while DTT totally inhibited the esterase activity. From the deduced amino acid sequence, it is suggested that the esterase harbors a catalytic triad consisting of Ser, His, and Asp. In order to confirm this experimentally, the effect of typical serine modifying reagents was tested. PCMB reduced the activity down to 30% whereas phenylmethylsulfonyl fluoride (PMSF) drastically inhibited the esterase activity. Inhibitors such as Pefablock (serine protease) and 2-iodoacetate (cysteine protease) also drastically reduced activity. The recombinant esterase activity was totally inhibited in the presence of EDTA (ethylenediamine tetra-acetic acid).

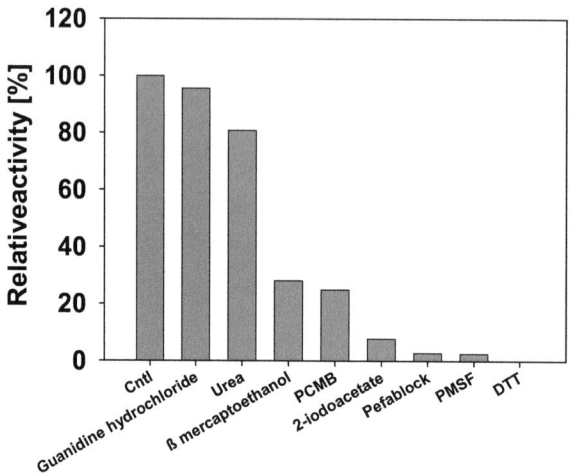

Fig. 24 Effect of various inhibitors on esterase activity was examined using Assay A (s. 2.3.1). The esterase was incubated with various inhibitors (10mM) for 1 h at 30°C. For examination of the enzyme residual activity, 100 μl of preincubation mixture were mixed with 900 μl of substrate (PNP butyrate) in 25 mM Tris-HCl buffer pH 7.0. The reaction was carried out for 20 min at 20°C, 1100 rpm.

The activity of the esterase was tested after incubation for 1 hour at 20°C in the presence of different detergents at a concentration of 10 % w/v (Fig. 25). CHAPS (3-[(3-chol-amido-propyl) dimethylammonio]-1-propanesulfonic acid) has a mild effect on the esterase activity. Incubation with Tween-80 or Triton X-100 decreased the activity down by 50%. SDS and tween 20 completely inhibited enzyme activity.

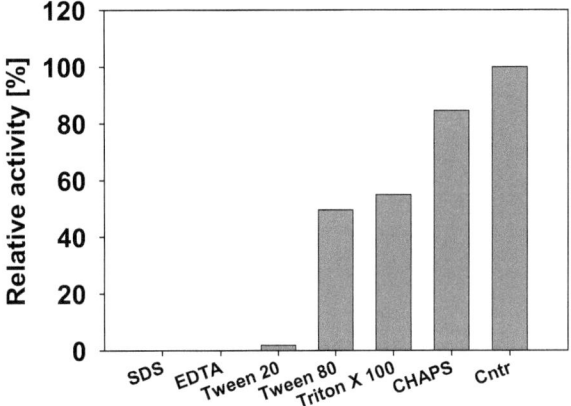

Fig. 25 Effect of different detergents on esterase activity was examined using Assay A (s. 2.3.1). The recombinant esterase was incubated with detergents (10% w/v) for 60 min at 20°C. For examination of the enzyme residual activity, 100 µl of preincubation mixture was mixed with 900 µl of substrate (PNP butyrate) in 25 mM Tris-HCl buffer pH 7.0. The reaction was carried out for 20 min at 20°C, 1100 rpm.

The effect of organic solvents is shown in Fig. 26. Most organic solvents decreased the activity of the recombinant esterase. The best solvent was 49% DMSO where the esterase showed 60% of the original activity. The esterase showed up to 40% of the original activity in 98 % Hexadecane and n-hexane. Activity was totally inhibited by 98 % amylalcohol, isopropanol, pyridine, and 49% ethanol.

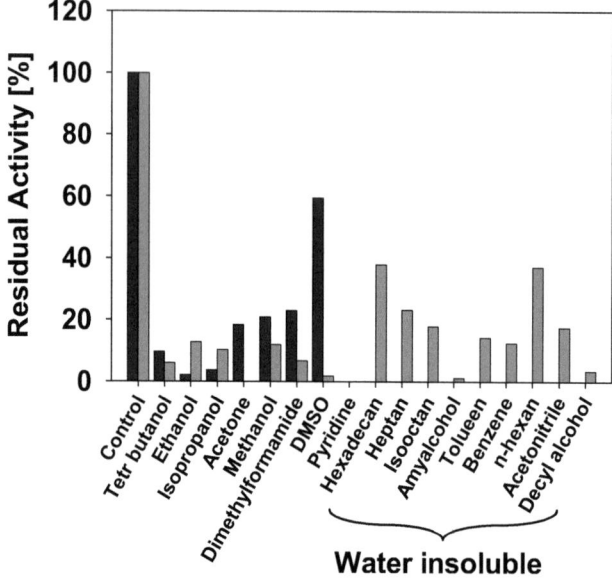

Fig. 26 Effect of various organic solvents on recombinant esterase activity was examined using Assay A (s. 2.3.1). The esterase was incubated with various organic solvents (49% and 98%) for 60 min at 20°C. For examination of the enzyme residual activity, 100 µl of preincubation mixture was mixed with 900 µl of substrate (PNP butyrate) in 25 mM Tris-HCl buffer pH 7.0. The reaction was carried out for 20 min at 20°C, 1100 rpm. ▉ Grey (98%), ▉ Black (49%)

3.2.7 Substrate specificity and kinetic parameters

Specificity of the enzyme towards the length of different acyl chains of pNP-Esters was investigated (1 mM in 25 mM Tris buffer; pH 7.0) (Fig. 27). The activity of the esterase is high in the presence of short-chain length C2-C8 pNP-Esters. p-nitrophenyl butyrate (C4) was the best substrate among the p-nitrophenyl esters examined. The recombinant esterase exhibited very low or no activity toward the long-chain (C12 - 18) substrates.

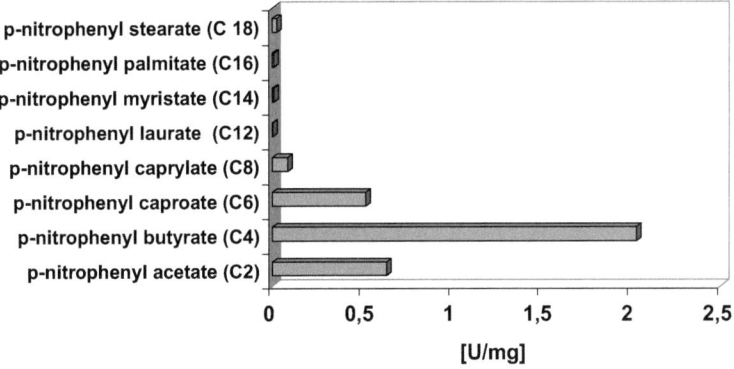

Fig. 27 Substrate specificity of the recombinant esterase using Assay A.
The enzyme assay was performed with 1 mM *p*NP-esters as substrates
(Sigma®) for 20 min at 20°C in 25 mM Tris-HCl buffer pH 7.0

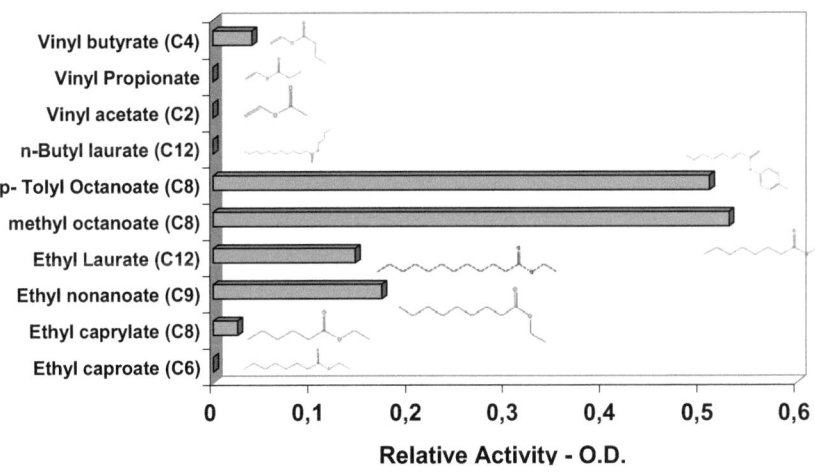

Fig. 28 Substrate specificity of the recombinant esterase using Assay B.
The enzyme assay was performed with 0.25 mg/ml substrate (Fluka®)
in 25 mM Tris buffer (pH 7) for 12 hours

Specificity of the enzyme towards the length of different acyl chains of ethyl,
methyl and vinyl esters was investigated (0.25mg/ml 25 mM Tris buffer; pH 7.0)
(Fig. 28). The activity of the esterase is high in the presence of long-chain length

C8-C12 ethyl esters. Ethyl nanoate (C9) was the best substrate among the ethyl esters examined. The esterase showed high activity toward methyl octanoate (C8) and good activity toward vinyl butyrate (C4) (Fig. 28).

Several substrates were used to test positional specificity and ability to hydrolyze molecules with chiral carbons (Fig. 29, 30). The esterase was able to hydrolyze ester bonds at the (1) position (4-nitrophenyl-1-naphtoate) rather than at the (2) position (4-nitrophenyl-2-naphtoate) (Fig. 29). The esterase was active with plenary molecules such as 4-nitrophenyl benzoate (Fig. 28). When tested for its ability to hydrolyze amide bonds in molecules such as 2-(4-isobutylphenyl)-N-(4-nitrophenyl) propanamide, no activity was detected (Fig. 29).

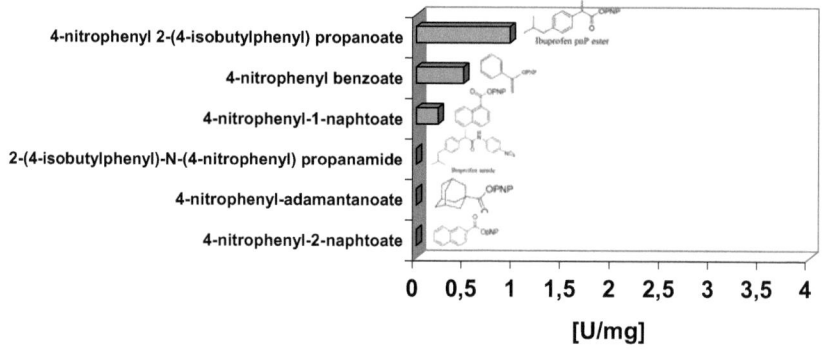

Fig. 29 Substrate specificity of the recombinant esterase using Assay A. The enzyme assay was performed with 0.25 mg/ml substrate in 25 mM Tris buffer (pH 7.5) for 30 min at 20°C.

The recombinant esterase was able to hydrolyze several ester bonds in molecules bearing chiral carbons, such as 4-nitrophenyl 2-(4-isobutylphenyl) propanoate, 4-nitrophenyl 2-phenylpropanoate, (S)-4-nitrophenyl 2-(6-methoxynaphthalen-2-yl) propanoate, and 4-nitrophenyl 3-phenylbutanoate (Fig. 30).

Finally, triacylglycerols such as triolein and tributyrate were inert as substrates. Tributyrin showed activity on agar plates (clearing zones) upon prolonged incubation; however it showed no activity in the spectrophotometric assay.

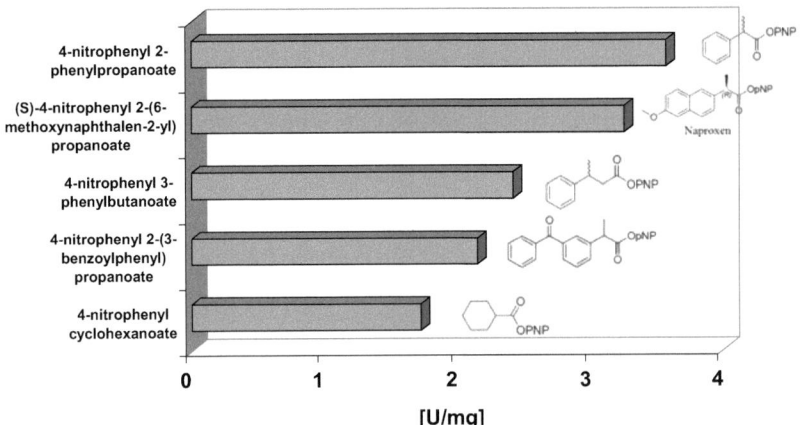

Fig. 30 Substrate specificity of the recombinant esterase using Assay A. The enzyme assay was performed with 0.25 mg/ml substrate in 25 mM Tris buffer (pH 7.5) for 30 min at 20°C.

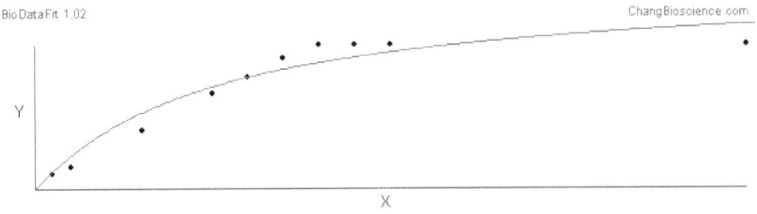

Fig. 31 Michaelis-Menten-Kinetics of recombinant esterase using the BioDataFit 1.02 program. Effect of substrate concentration on the velocity of enzyme. In order to calculate the k_m and V_{max}, the enzyme activity was tested using method A at 20 °C, pH 7 and for 30 min with different concentrations of p-nitrophenyl butyrate (0.05 – 2 mM). For each reaction 2 µl (2.0 U/mg) enzyme solution was used. One unit of esterase is defined as the amount of enzyme that releases 1 µmol of p-nitrophenol per min under assay conditions specified.

The apparent kinetic parameters for Michaelis-Menten- with different concentrations of p-nitrophenyl butyrate $(0.05 - 2$ mM) were determined at pH 7 at 20 °C. The K_m value for esterase is 0.511 mM and, V_{max} value is 5.59 U/mg as shown in Fig. 31.

$V0 = VMax [S] / (KM + [S])$.

Results:

Model: a x / (x + b)

V max = a = 5.59 U/mg

Km = b = 0.511 mM

EC50 = 0.511726

Variance Analysis

Source	DF	Sum of Squares	Mean Square
Regression	1	16.04968	16.04968
Error	8	1.137350	0.142168
Total	9	17.18703	

Standard Error of the Estimate: 0.377052

R-squared: 0.933825

C h a p t e r   4

4   DISCUSSION

4.1   *Pseudoalteromonas arctica* sp. nov., a novel psychrotolerant bacterium isolated from Spitzbergen

Marine *Pseudomonas*-like bacteria comprise several related genera, including *Alteromona*, *Marinomonas*, *Pseudoalteromonas*, *Moritella*, *Marinobacter*, *Psychrobacter*, *Colwellia*, *Shewanella* and *Halomonas*. Classification of these aerobic heterotrophic micro-organisms from the marine environment remains difficult and time-consuming due to their phenotypic diversity and absence of robust chemotaxonomic markers. Fortunately, phylogenetic information, based on 16S rRNA gene sequences, is useful in classifying and identifying micro-organisms belonging to previously poorly defined taxa (Ivanova *et al.*, 2000).

The genus *Pseudoalteromonas*, originally called *Alteromonas*, included non-pigmented, Gram-negative, heterotrophic, aerobic, polarly flagellated species of marine bacteria which had G+C contents ranging from 38 to 50 mol% which differentiated this bacterial group from the previously described genus *Pseudomonas* (55 to 64 mol%). Since Baumann et al. (Baumann *et al.*, 1972) created the genus *Alteromonas* in 1972, several species have been assigned to this genus, although a recent revision of genera based on phylogenetic analysis by Gauthier et al. (Gauthier *et al.*, 1995) divided the genus *Alteromonas* into two genera *Pseudoalteromonas* and *Alteromonas*. Separation of *Pseudoalteromonas* species on the basis of phenotypic characteristics is problematic because of significant variations in phenotypic traits and because phenotypic differences have frequently been observed even among genetically closely related strains. Thus, in order to determine the relationships between *Pseudoalteromonas* species, genetic and chemotaxonomic methods appear to provide more reliable information than differential phenotypic characteristics (Bozal *et al.*, 1997).

Pseudoalteromonas is one of the largest genera within the γ-Proteobacteria and currently comprises more than 30 species (Ivanova *et al.*, 2004). Bacteria of the genus *Pseudoalteromonas* have been the objects of intensive studies during the last three decades. These bacteria play an important role in marine environments owing to their abundance and high metabolic activities. Pseudoalteromonads are highly capable of surviving in nutrient-poor marine environment by adjustment of their biochemical pathways and production of a wide variety of metabolites, including biologically active compounds and enzymes (Ivanova *et al.*, 2003).

The strain A 37-1-2 showed high similarity regarding it`s morphological characterization to the strains belonging to the genus *Pseudoalteromonas* identified as: Gram negative, polarly flagellated, non spore forming, motile short rods (0.15-0.2 μm wide and 0.4-0.75 μm long) (Bozal *et al.*, 1997; Ivanova *et al.*, 2000; Ivanova *et al.*, 2004; Romanenko *et al.*, 2003; Sawabe *et al.*, 2000).

The optimal growth temperature and pH of the strain A 37-1-2 (10–15 °C and pH 7–8) were similar to related *Pseudoalteromonas* species: *Pseudoalteromonas antarctica* (10-15 °C and pH 7-8), *Pseudoalteromonas aliena* (22-25 °C and pH 7-8), *Pseudoalteromonas distincta* (20-25 °C and pH 7-8), and *Pseudoalteromonas elyakovii* (25-30°C and pH 7-8). The strain A 37-1-2 grew at 4°C which is similar to strains *Pseudoalteromonas espejiana*, *Pseudoalteromonas carrageenovora*, and *Pseudoalteromonas haloplanktis* that can grow at 4°C as well. The strain A 37-1-2 grew in the presence of a wide range of NaCl concentration, from 0 to 9 % (w/v) (optimum 2 to 3 % (w/v)) which is similar to the range for the growth of *Pseudoalteromonas distincta* (0.5-8%) and *Pseudoalteromonas aliena* (0.5-6%), (Bozal *et al.*, 1997; Ivanova *et al.*, 2004).

The strain A 37-1-2 is able to utilize a large number of different substrates: α-Cyclodextrin, Dextrin, D-Cellobiose, D-Galactose, α-D-Glucose, Maltose, D-Mannitol- D-Melibiose, Sucrose, Pyruvic acid methyl ester, D- Gluconic acid, L-Alanine, L-Alanylglycine, Glycyl-L-Glutamic acid, L-Serine, Esculine ferric citrate, and Capric acid. *Pseudoalteromonas elyakovii*, *Pseudoalteromonas espejiana* *Pseudoalteromonas citrea*, *Pseudoalteromonas macleodii* and the strain A 37-1-2 shared the ability to utilize: D- galactose, sucrose, maltose, Melibiose, D-mannitol, and pyruvate. *Pseudoalteromonas elyakovii*, *Pseudoalteromonas haloplanktis*, *Pseudoalteromonas macleodii* and strain A 37-1-2 shared the ability to utilize D-gluconate.

The strain A 37-1-2 produced several enzymes: pullulanase, protease, β-Glucanase, esterase, pectinase, β-glucosidase, and β-galactosidase. *Pseudoalteromonas aliena*, *Pseudoalteromonas antarctica*, and *Pseudoalteromonas agarivorans* are known to produce proteases (Bozal *et al.*, 1997; Ivanova *et al.*, 2004; Romanenko *et al.*, 2003). *Pseudoalteromonas agarovorans* produces an esterase and a β-galactosidase (Bozal *et al.*, 1997).

The strain A 37-1-2 has 18.08 % straight chain saturated FAME, 9.61 % terminally branched saturated FAME and 71.45% monounsaturated FAME. 16:0 straight chain saturated FAME (8.23 %) and 16:1 w7c monounsaturated FAME (58.60 %) were the most abundant FAME s found in the strain A 37-1-2 as well as, in the related strains: *Pseudomonas haloplanktis* (Bozal *et al.*, 1997), *Pseudoalteromonas atlantica* (Bozal *et al.*, 1997), *Pseudoalteromonas aliena* (Ivanova *et al.*, 2004), *Pseudoalteromonas agarovorans* (Romanenko *et al.*, 2003) and *Pseudoalteromonas antarctica* (Bozal *et al.*, 1997).

High similarity was shown between the G + C content of strain A 37-1-2 (39 mol %) and related strains: *Pseudoalteromonas elyakovii* (38.5, 38.9%), *Pseudoalteromonas distincta* (43.6%), *Pseudoalteromonas aliena* (41, 43%), *Pseudoalteromonas agarovorans* (42.2, 43.8%), *Pseudoalteromonas antarctica* (41, 42%)

and *Pseudoalteromonas carrageenovora* (39.5%) (Bozal *et al.*, 1997; Ivanova *et al.*, 2000; Ivanova *et al.*, 2004; Romanenko *et al.*, 2003). Phylogenetic analysis placed the strain A 37-1-2 (according to its 16S rDNA composition) among the species of the genus *Pseudoalteromonas* (about 30 species). The strain A 37-1-2 shared (according to its 16S rDNA composition) 99 % identity with *Pseudoalteromonas elyakovii* (identities: 1484/1489), *Pseudoalteromonas citrea* (identities: 1469/1474), *Pseudoalteromonas haloplanktis* (identities: 1484/1494), *Pseudoalteromonas distincta* (identities: 1473/1481), *Pseudoalteromonas agarovorans* (identities: 1475/1484) and *Pseudoalteromonas gracilis* (identities: 1484/1494). Although the level of 16S rRNA similarity with the closest relative (99%) is high for two species, the observed value is still low for the average level of similarity within the genus (Gauthier *et al.*, 1995). The strain A 37-1-2 formed a cluster with *Pseudoalteromonas elyakovii* and *Pseudoalteromonas distincta*. DNA-DNA hybridization of the strain A 37-1-2 and the most identical strain *Pseudoalteromonas elyakovii* showed only 47.6 % DNA–DNA similarity despite the fact that the 16S rDNA identity between both, was very high (99 %). This strongly indicates that the newly isolated strain A 37-1-2 is a new species within the genus *Pseudoalteromonas*.

All data mentioned previously demonstrate clearly that the new isolate A 37-1-2 is a new species within the genus *Pseudoalteromonas*. This fact was also supported by the DNA-DNA hybridization, phylogenetic, G+C content, morphological and fatty acid analysis data. Based on the phylogenetic analysis by 16S rDNA and DNA-DNA hybridization homology, the new isolated strain A 37-1-2 was found to be closely related to *Pseudoalteromonas elyakovii*, but represents a new species within the genus *Pseudoalteromonas*. We propose to assign the new isolate to the genus *Pseudoalteromonas* as *Pseudoalteromonas arctica* sp. nov.

4.2 Esterases from psychrophilic aerobic bacteria

Many microorganisms are known to produce esterases. Recently, reports on a few microbial, c
Acinetobacter sp. (Suzuki *et al.*, 2002), *Pseudomonas* sp. (Suzuki *et al.*, 2003), and
Psychrobacter sp. (Kulakova *et al.*, 2004) have been cloned and characterized.

This report presents the cloning, purification and properties of a cold-active
esterase from a novel psychrotolerant marine isolate *Pseudoalteromonas arctica* sp.
nov.

4.3 Cloning and expression of an esterase gene from *Pseudoalteromonas arctica* sp. nov. in *E.coli*

The *Pseudolateromonas arctica* gene (*Est 37*) encoding an esterase (Est 37) was
cloned and sequenced. Multi-sequence alignment of the amino acid sequence of
this esterase with the most homologous proteins indicates at least five regions
with a high degree of identity (Fig. 32).

The comparison with the α/β-hydrolase sequences predicted the structure of the
α/β-hydrolase fold for the esterase from *Pseudoalteromonas arctica* (Est 37). The
α/β-hydrolase fold family of enzymes is one of the largest groups of structurally
related enzymes with diverse catalytic functions. Members in this family include
acetylcholinesterase, dienelactone hydrolase, lipase, thioesterase, serine
carboxypeptidase, proline iminopeptidase, proline oligopeptidase, haloalkane
dehalogenase, haloperoxidase, epoxide hydrolase, hydroxynitrile lyase and others
(Holmquist, 2000). These enzymes have a catalytic triad evolved to efficiently
operate on substrates with different chemical composition or physicochemical
properties and in various biological contexts. A catalytic triad of the Est 37
consisting of one serine, one aspartic acid residue and one histidine was
predicted (Brady *et al.*, 1990). It is therefore proposed that Est 37 belongs to the
serine hydrolase family with a potential catalytic triad comprising Ser106, found

within the G-x-S-x-G consensus sequence, known as the substrate-binding site from many esterases (Brenner, 1988), Asp196 and His225. Finally, the binding site termed the oxyanion hole comprising a His33 was predicted.

Box I   Box II   Box III

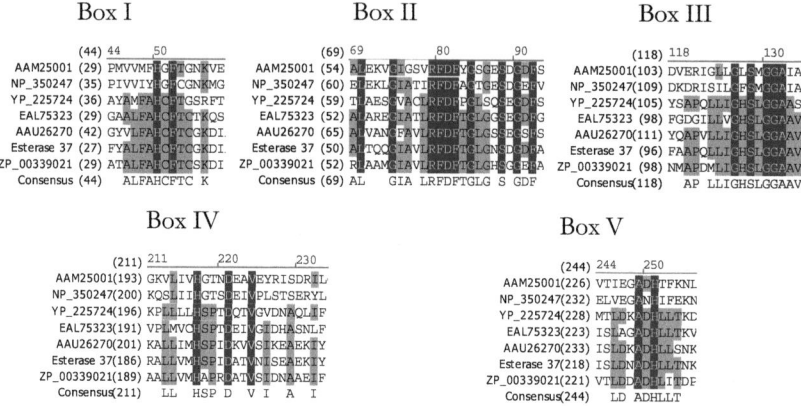

Box IV   Box V

Fig. 32 Comparison of microbial α/β hydrolase sequences. The alignment of the deduced amino acid sequence of *Pseudoalteromonas arctica* Est 37 and homologous α/β hydrolases. The accession numbers of the aligned sequences are for the following organisms: AAM25001, α/β hydrolase from *Thermoanaerobacter tencongensis* MB4; NP_350247, α/β hydrolase from *Clostridium acetobutylicum* ATCC 824; YP_225724, α/β hydrolase from *Corynebacterium glutamicum* ATCC 13032; EAL75323, α/β hydrolase from *Erythrobacter litoralis* HTCC 2594; AAU26270, α/β hydrolase from *Legionella pneumophila* subsp. *pneumophila* str. Philadephia 1; ZP_0033902, α/β hydrolase from *Silicibacter* sp. TM 1040. The accession numbers are indicated to the left of the amino acid sequences. Identical residues have a black background and similar residues have a grey background.

At present, lipolytic enzymes from prokaryotes are classified into eight families (I to VIII) according to their amino acid sequences (Arpigny & Jaeger, 1999). Computer analysis of the deduced amino acid sequence of the *Pseudoalteromonas arctica* esterase established that it does not belong to any of the described families. Est 37 bears an amino acid sequence with the following conserved blocks: Block I - HCFT, Block II- RFDF and Block III – GxSxGGA (Fig. 32).

The *Pseudoalteromonas arctica* esterase showed no significant similarity with most identified esterases (Table 10). At the amino acid level it shares less than 10% similarity with any known esterase gene. The highest identity was obtained with three bacterial esterases: 10% with the polyurethane esterase (BAA76305) from the soil bacterium *Delftia acidovorans*/ATCC15668, 9% with a lipolytic enzyme (AAC67392) from *Sulfolobus acidocaldarius*/ATCC33909 and 9% with an esterase (BAA25795) from *Acetobacter pasteurianus*/LMD 51.1.

The genes of cold-adapted esterases from psychrotrophic bacteria *Psychrobacter* sp. Ant300 (Kulakova *et al.*, 2004), *Pseudomonas* sp. B11-1 (Suzuki *et al.*, 2003) and *Acinetobacter* sp. strain No.6 (Suzuki *et al.*, 2002) isolated in Antarctica, Alaska and Siberia respectively were cloned and sequenced. The cold-active enzymes showed high activities at temperatures as low as 5°C. These esterases are active at low temperatures and are not thermostable. The composition of amino acids of the esterases from the psychrotolerant strain *Pseudoalteromonas arctica*, the psychrophilic strains *Psychrobacter* sp. Ant300, and *Acinetobacter* sp. strain No.6, the thermophilic bacterial strain *Bacillus licheniformis* (Alvarez-Macarie *et al.*, 1999) and the mesophilic bacterial strain *Lactobacillus casei* LILA (Fenster *et al.*, 2003) are presented in Table 14.

Cold and thermo-active enzymes can be differentiated only by the temperature ranges in which enzymes are stable and active. Otherwise, they tend to be highly similar. The sequences of homologous thermophilic and mesophilic proteins are typically 40 to 85% similar (Vieille & Zeikus, 2001). The three-dimensional structures of cold and thermo-active enzymes are superposable and they have the same catalytic mechanisms (Vieille & Zeikus, 2001). A highly flexible structure of cold-active enzymes can provide enhanced ability to undergo discrete and fast conformational changes at low temperatures imposed by the catalytic events (Feller & Gerday, 1997). The conformation of the cold-adapted enzymes is less compact at all temperatures in comparison to that of the thermo-adapted enzymes. Hyperthermophilic enzymes are more rigid than their mesophilic homologues at mesophilic temperatures and that rigidity is a prerequisite for high protein thermostability (Vieille & Zeikus, 2001).

Table 14 Amino acid composition of cold-active and thermophilic esterases.

Amino Acid(s)	P. arctica[a]		Psychrobacter[b]		Acinetobacter[c]		Lactobacillus[d]		Bacillus[e]	
	No*	%**	No	%	No	%	No	%	No	%
Charged (RKHYCDE)	131	32.75	119	29.75	92	30.56	88	33.98	137	28.31
Acidic (DE)	51	12.75	44	11.00	26	8.64	34	13.13	64	13.22
Basic (KR)	43	10.75	30	7.50	31	10.30	27	10.42	46	9.50
Polar (NCQSTY)	96	24.00	100	25.00	87	28.90	66	25.48	99	20.45
Hydrophobic (AILFWV)	150	37.50	147	36.75	108	35.88	87	33.59	176	36.36
A Ala	47	11.75	39	9.75	24	7.97	17	6.56	44	9.09
C Cys	6	1.50	10	2.50	4	1.33	3	1.16	3	0.62
D Asp	29	7.25	28	7.00	19	6.31	21	8.11	26	5.37
E Glu	22	5.50	16	4.00	7	2.33	13	5.02	38	7.85
F Phe	14	3.50	11	2.75	15	4.98	9	3.47	23	4.75
G Gly	23	5.75	30	7.50	17	5.65	16	6.18	39	8.06
H His	21	5.25	21	5.25	12	3.99	13	5.02	16	3.31
I Ile	24	6.00	22	5.50	18	5.98	20	7.72	27	5.58
K Lys	28	7.00	19	4.75	15	4.98	15	5.79	27	5.58
L Leu	41	10.25	46	11.50	33	10.96	25	9.65	48	9.92
M Met	4	1.00	7	1.75	6	1.99	2	0.77	14	2.89
N Asn	20	5.00	12	3.00	15	4.98	13	5.02	16	3.31
P Pro	12	3.00	21	5.25	14	4.65	14	5.41	30	6.20
Q Gln	13	3.25	17	4.25	18	5.98	5	1.93	18	3.72
R Arg	15	3.75	11	2.75	16	5.32	12	4.63	19	3.93
S Ser	22	5.50	24	6.00	21	6.98	17	6.56	31	6.40
T Thr	25	6.25	23	5.75	10	3.32	17	6.56	23	4.75
V Val	22	5.50	24	6.00	17	5.65	15	5.79	26	5.37
W Trp	2	0.50	5	1.25	1	0.33	1	0.39	8	1.65
Y Tyr	10	2.50	14	3.50	19	6.31	11	4.25	8	1.65

*Esterases from: [a] *Pseudoalteromonas arctica*, [b] *Psychrobacter* sp. Ant300, [c] *Acinetobacter* sp. strain No.6, [d] *Lactobacillus casei* LILA, [e] *Bacillus licheniformis*. * Number count, ** % frequency.

The numerous homology-based models and especially the recently solved X-ray structures of cold-active enzymes have deciphered the molecular origin of the weak stability characterizing these proteins. All structural factors currently known to stabilize the protein molecule can be attenuated in strength and number in the structure of cold-active enzymes. This involves the clustering of glycine residues (providing local mobility), the disappearance of proline residues in loops (providing enhanced chain flexibility between secondary structures), a reduction in arginine residues which are capable of forming multiple salt bridges and H bonds, as well as a lower number of ion pairs, aromatic interactions or H bonds compared to mesophilic enzymes (Feller, 2003).

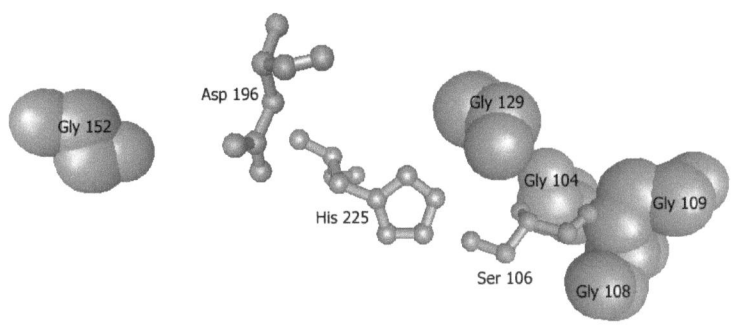

Fig. 33 A model of the catalytic site of Est 37 built using the 3D-JIGSAW comparative modeling server (http://www.bmm.icnet.uk/~3djigsaw/), showing Glycine molecules clustering around the catalytic site. This model was constructed based on the X-ray crystallographic structure of a *Pseudomonas fluorescens* aryl esterase (Cheeseman *et al.*, 2004).

A three dimentional structural model for the esterase domain of Est 37 was constructed on the basis of the 2.5-Å crystal structure of *Pseudomonas fluorescens* aryl esterase (Cheeseman *et al.*, 2004) with an appreciable sequence similarity (15% identity) to the esterase domain of Est 37 (Fig. 33). This showed that a cluster of 5 glycine molecules surrounds the catalytic site of Est 37 (Fig. 33) thus providing local mobility at the active site and contributing to its cold- catalytic activity.

A low frequency of occurrence of proline molecules was observed in Est 37 (3.00%) and the cold-active esterases (4.65 and 5.25%) as compared to the mesophilic (5.41%) and thermo-active esterases (6.20%). The same was also noted in the case of arginine were a lower frequency of 2.75 and 3.75% was observed in cold-active esterases as compared to 4.63% in the mesophilic esterase (Table 14).

In cold-active enzymes the size and relative hydrophobicity of non-polar residue clusters forming the protein core are frequently smaller, lowering the compactness of the protein interior. The N and C caps of a helices are also altered (weakening the charge-dipole interaction) and loose or relaxed protein extremities appear to be preferential sites for unzipping. The binding of stabilizing ions, such as calcium, can be extremely weak, with binding constants differing from mesophiles by several orders of magnitude. Insertions and deletions are sometimes responsible for specific properties such as the acquisition of extra surface charges (insertion) or the weakening of subunit interactions (deletion). Calculation of the solvent-accessible area showed that some cold-active enzymes expose a higher proportion of non-polar residues to the surrounding medium. This is an entropy-driven destabilizing factor caused by the reorganization of water molecules around exposed hydrophobic side chains. Calculations of the electrostatic potential revealed in some cases an excess of negative charges at the surface of the molecule and, indeed, the pI of cold-active enzymes is frequently more acidic than that of their mesophilic or thermo-active homologues. This has been related to improved interactions with the solvent, which could be of prime importance in the acquisition of flexibility near zero degrees (Feller, 2003). The esterases from *Pseudoalteromonas* and *Psychrobacter* have negative charges of -6.41 and -12.46 respectively at pH 7.

Besides the balance of charges, the number of salt bridges covering the protein surface is also reduced. There is now a clear correlation between surface ion pairs and temperature adaptation, since these weak interactions significantly increase in number from psychrophiles to mesophiles, thermophiles and hyperthermophiles, the latter showing arginine-mediated multiple ion pairs and interconnected salt bridge networks. Such an altered pattern of electrostatic interactions at the molecular surface is thought to improve the resilience or the 'breathing' of the external shell of cold-active enzymes. Of course, all these factors are not found in every cold-active enzyme: each enzyme adopts its own strategy by using one or a combination of these altered structural factors to improve the local or global mobility of the protein edifice (Feller, 2003).

Putative conserved domains of the Est 37 have been detected using the EBI Pro Scan. This Esterase has an N-terminal esterase domain and an a C-terminal OsmC (Osmotic shock response protein) domain (Fig.14). Non-sporulating bacteria have evolved a number of mechanisms to deal with environmental stress including that of heat, oxidative agents and osmotic shock. The response to different types of stress is complex, involving the overlapping expression of multiple genes. The expression of the OsmC gene was first demonstrated in *Escherichia coli* in response to osmotic shock (Gutierrez & Devedjian, 1991). The resultant protein is therefore referred to as osmotically inducible protein C. The distribution of the gene was studied by Völker et al (Volker *et al.*, 1998) and was found in a number of bacterial species as one or two copies. Although the function of the protein remains unknown, the gene was found to be highly conserved (Rehse *et al.*, 2004).

Exported proteins are synthesized initially as preproteins with an aminoterminal extension. This signal peptide, which distinguishes the secreted proteins from cytoplasmic ones, is needed for targeting to the export pathway (Simonen & Palva, 1993). The *Pseudoalteromonas arctica* esterase does not have an apparent signal sequence indicating that the enzyme is intracellular.

The esterase gene from *Pseudoalteromonas arctica* was cloned and successfully expressed in *E. coli*. The expression was high comprising up to 30% of the total cell protein. The recombinant esterase was not secreted to the culture supernatant. High production obtained of the recombinant esterase (138 U/liter culture and 79.5 mg/1 liter culture) makes this method a very good potential source of this unique enzyme.

4.4   Purification and characterization of the recombinant esterase

The main constraints in traditional purification strategies include low yields and long time purification periods. For industrial purposes, the purification strategies employed should be inexpensive, high yielding, rapid and amenable to large-scale operations. Most of the commercial applications of enzymes do not always need homogeneous preparations. A certain degree of purity is required, depending upon the final application, and should enable efficient and successful usage. In this case, the protein purification in a single step on a Ni column is a suitable, inexpensive and an efficient strategy.

The recombinant esterase was purified by a single purification step and characterized. After Ni-NTA superflow (Qiagen) Column chromatography the recombinant esterase was purified 3.3 Fold, with a specific activity 1.7 U/mg, and a yield of 96 %. The purified esterase has a specific activity of 1.7 U/mg which is relatively low when compared to other known cold-active esterases like 58.2 U/mg for the esterase from *Pseudomonas* sp. Strain B11-1 (Suzuki *et al.*, 2003) and 115 U/mg for the esterase from *Acinetobacter* sp. Strain No. 6 (Suzuki *et al.*, 2002).

The maximal temperature and pH under standard assay conditions are 25°C and pH 7.5. Up to 50% of the activity is retained at 0 °C. Similar cold-adapted esterases showed varying optima like (35°C, pH 7.9) for the esterase from *Psychrobacter* sp. Ant300, (45°C, pH 7.5) for the esterase from *Acinetobacter* sp. Strain No.6, and (45°C, pH 8.0) for the esterase from *Pseudomonas* sp. strain B11-1.   However none of these esterases showed similar high activity at low

96

temperatures. The relative activities at 5°C of the esterases from *Psychrobacter* sp. Ant300, *Acinetobacter* sp. Strain No.6, and *Pseudomonas* sp. strain B11-1 were 10%, 12%, and 20% respectively. The activation energy E_a for the Est 37 catalyzed hydrolysis of pNP-butyrate was calculated from Arrhenius plot to be 5.1 kcal/mol in the temperature range from 0 to 25°C. This value was lower than the values reported for the same reaction catalyzed by mesophilic lipases/esterases: 28.4 kcal/mol for lipase PS of *Pseudomonas* sp. and 21.0 kcal/mol for lipase from *Candida cylindracea* (Suzuki *et al.*, 2002). Moreover, the value was even lower than the values reported for cold-active esterases and lipases from other sources: *Pseudomonas* sp. strain B11-1, 11.2 kcal/mol for LipP (Choo *et al.*, 1998) and 20.1 kcal/mol for PsEst1 (Suzuki *et al.*, 2003) *Acinetobacter* sp. strain No.6, 9.0 kcal/mol for AEST (Suzuki *et al.*, 2002) and 11.2 kcal/mol for AELH (Suzuki *et al.*, 2002). This low activation energy was matched only by that of a cold-active esterase, (PsyEst) 4.6 kcal/mol, from *Psychrobacter* sp. Ant 300. The observed low activation energy was consistent with the general observation that the activation energies of reactions catalyzed by cold-active enzymes are lower than those catalyzed by their mesophilic counterparts.

The recombinant esterase is stable at temperatures of 4 and 10 °C. It lost around 35% activity after 24 hours incubation at 20°C. It has a half life of 15 hours at 30 °C and 5 hours at 40°C. Moreover, Est 37 is stable at alkaline rather than at acidic pHs. Est 37 is thus more thermostable than most known cold active esterases such as PsyEst from *Psychrobacter* sp. Ant300 (Kulakova *et al.*, 2004), PsEst1 from *Pseudomonas* sp. Strain B11-1, and AEST from *Acinetobacter* sp. Strain No. 6 which have half lifes of 16min, 15min and 60 min respectively at 40°C.

A hydrolase-catalyzed reaction in a conventional aqueous system thermodynamically favors hydrolysis. Because an esterase-catalyzed reaction is also such a case, a condensation reaction to produce an ester is carried out in a non-aqueous medium such as an organic solvent, a solvent-free system, or an ionic liquid. To carry out a condensation reaction the esterase should be stable in the organic solvent. The organic solvents recommended for Est 37 are 50% DMSO and 100% hexadecane where it showed 60% and 40% stability respectively upon incubation for 60 min. Most other organic solvents decreased the activity of the recombinant esterase. Activity was totally inhibited by 100% amylalcohol, isopropanol, pyridine, and 50% ethanol.

Generally, esterases are cofactor-independent enzymes. Est 37 was inhibited by 10 mM of Al^{+2} (100% Inhibition), Cu^{+2} (100% Inhibition), Fe^{+2} (100% Inhibition), Ni^{+2}, Cr^{+2}, and Co^{+2}; whereas, Mg^{+2}, Se^{+2}, Mn^{+2} and K^{+1} had no or limited effect. Na^{+1} slightly increased the activity by 30%. Cu^{+2} and Fe^{+2} were reported to inhibit activity of the esterase AEST from *Acinetobacter* sp. Strain No. 6 (Suzuki *et al.*, 2002), and esterase 14468 from *Mycobacterium smegmatis* (Tomioka, 1983). Ca^{+2} is known to increase activity in some lipases and was reported to slightly increase the activity of PsyEst from Psychrobacter sp. Ant300 (Kulakova *et al.*, 2004) whereas Ca^{+2} slightly inhibited the activity of Est 37.

Various inhibitors were used to study the structure and catalytic mechanism of the esterase. There are two classes of enzyme inhibitors reversible and irreversible. The reversible inhibitors are classified further in two groups as nonspecific and specific reversible inhibitors. Compounds that inhibit esterase activity by changing the conformation or interfacial properties, but do not act directly on the active site are defined as non-specific inhibitors such as surfactants. Specific inhibitors directly interact with the active site of the enzyme and can be either reversible or irreversible. Specific irreversible inhibitors inhibit the catalytic activity of enzymes by reacting with the amino acids at or near the active site. Serine inhibitors are potential esterase active-site irreversible inhibitors, because esterases belong to a class of serine hydrolases with the catalytic triad Ser-His-Asp. Phenylmethylsulfonyl fluoride (PMSF), phenylboronic acid, Pefablock and diethyl p-nitrophenyl phosphate belong to

98

serine-specific inhibitors. Pefablock and PMSF inhibited the activity of Est 37 completely and irreversibly which is consistent with Est 37 being structurally related to the serine hydrolase superfamily of enzymes. PMSF is also a potent inhibitor of many esterases such as the cold-active esterases PsyEst (Kulakova *et al.*, 2004) and AEST (Suzuki *et al.*, 2002). The esterase PsEst1 activity was not affected by PMSF which suggests that the catalytic residues may be buried within the enzyme molecule.

The protein sequence of Est 37 contains 6 cysteine residues. Therefore, the specific irreversible inhibitors such as 2-mercaptoethanol, iodoacetate and DTT, which may disturb sulphhydryl bonds and thus modify the protein conformation, had an inhibitory effect on the esterase activity.

The inhibition of Est 37 activity by EDTA could probably be due to the influence on the interfacial area between the substrate and the enzyme. No such inhibitory effect was observed on the activity of PsyEst from *Psychrobacter* sp. Ant300 (Kulakova *et al.*, 2004), PsEst1 from *Pseudomonas* sp. Strain B11-1, and AEST from *Acinetobacter* sp. Strain No. 6.

Pseudoalteromonas arctica ´s esterase is active on a wide range of substrates. The recombinant enzyme Est 37 demonstrated a wide range of substrate specificity depending on the chain length of fatty acids. The recombinant esterase exhibited high activity towards water soluble substrates with short chains, such as p-nitrophenyl esters with fatty acid chain shorter that C8. Triacylglycerols such as triolein and tributyrate were inert as substrates therefore the enzyme can be classified as a carboxylesterase (Bornscheuer, 2002). A criterion has been proposed in which carboxylesterases are defined as enzymes that catalyze the hydrolysis of acylglycerols with short chains (<10 carbon atoms), while lipases are defined as enzymes that catalyze the hydrolysis of acylglycerols with long chains (≥10 carbon atoms) (Jaeger *et al.*, 1999). Est 37 showed high activity with methyl esters such as methyl octanoate. It hydrolyzes ester bonds of ethyl esters were it tends to hydrolyze longer (C8-C12) rather than shorter chains which is in contrary to the activity observed with p-nitrophenyl esters. This could be due to the fact that the smaller size of the ethyl group, compared to the nitrophenyl group, allows better interaction of the substrate of longer chain with the catalytic site of the esterase. Est 37 showed to be regio-selective. It was able to hydrolyze ester bonds at the (1) position (4-nitrophenyl-1-naphtoate) rather than at the (2) position (4-nitrophenyl-2-naphtoate). Est 37 was not able to hydrolyze amide bonds in molecules such as 2-(4-isobutylphenyl)-N-(4-nitrophenyl) propanamide. The recombinant esterase was able to hydrolyze several ester bonds in molecules bearing chiral carbons, such as 4-nitrophenyl 2-(4-isobutylphenyl) propanoate, 4-nitrophenyl 2-phenylpropanoate, (S)-4-nitrophenyl 2-(6-methoxynaphthalen-2-yl) propanoate, and 4-nitrophenyl 3-phenylbutanoate. This can be optimized and developed in the future, to make the esterase useful for the synthesis of optically pure compounds. Several carboxylesterases such as AGE, esterase from *A. globiformis*; BCE, esterase from *Bacillus coagulans*; BSE, esterase from *Bacillus stearothermophilus*; BGE, esterase from *B. gladioli* ; PAE, esterase from *P. aeruginosa*; PME, esterase *from Pseudomonas marginata*; PPE, esterase from *Pseudomonas putida*; RRE, esterase from *Rhodococcus ruber*; SDE, esterase from *Streptomyces diastatochromogenes* are reported to catalyze the kinetic resolution of chiral compounds (Bornscheuer, 2002).

4.5   Possible applications of the recombinant esterase

Est 37 is clearly one of the most cold-active esterases known to date and this has been demonstrated by the very low activation energy needed for its catalytic activity. This "cold activity" (i.e., high catalytic activity at low temperatures) of Est 37 can be the key to the success in some of its applications. These applications include its use as additives in laundry detergents for cold washing and as a catalyst for organic syntheses of unstable compounds at low temperatures.

Knowing that esterases are important biocatalysts for the industrial production of chiral intermediates, Est 37 was able to hydrolyze several ester bonds in molecules bearing chiral carbons. Microbial esterases are very interesting for commercial processes because of good availability, high stability, and good catalytic characteristics. Chiral intermediates are attracting a growing interest as important intermediates for catalysts, liquid crystals, pharmaceuticals and agrochemicals. In general these products can be produced by three approaches: Traditional methods (classic resolution, chromatography, and chiral pool synthesis), asymmetric chemical methods, and biological methods. Biological methods include biocatalysis, biotransformation, and bioresolution. It is expected that the sales volume of chiral technologies will double in total until 2009. An extraordinary growth is expected for biological methods. Nevertheless, similar to chemical catalysts biological catalysts must also emphasize some critical parameters for industrial applications: High catalytic activity and selectivity, good stability, and availability for large-scale applications. Esterases are used for the selective hydrolysis of esters. Esterases have different advantageous characteristics if compared to other chemical or enzymatic catalysts. In general esterases are easy to handle. Reactions can be simply monitored and steered via pH-control. The enzymes are quite stable which is important in industrial processes. Hydrolysis reactions are performed under mild conditions. Additionally, esterases are characterized by a broad substrate spectrum combined with high stereoselectivities. A wide variety of applications have been described for this class of enzymes. The desymmetrization of prochiral substrates is a very efficient example. Analogous reactions lead to the corresponding chiral products with a theoretical yield of 100 %. In contrast to racemic resolution reactions, optical purity is not dependent on the rate of conversion. Esterases were also used for the optical resolution of racemates. For instance one enantiomer of α-substituted methyl esters is selectively hydrolyzed to the corresponding chiral acid whereas the enantiomeric ester remains unreacted (Jülich Chiral Solutions, http://www.juelich-chemicals.de/public/page-view.cfm?InfoID=59).

Est 37 showed a good potential in applications that require regioselectivity such as regioselective hydrolysis of an ester group in the presence of a second ester function. The simple inactivation of Est 37 by the rise of the temperature or by addition of the serine-specific inhibitors is a further advantage for the employment of the recombinant protein in industrial processes, where the enzyme can be stopped easily after the expiration of a desired reaction.

A major problem for the application of esterases is the rare commercial availability. Different enzymes such as pig liver esterase and horse liver esterase were isolated from animal sources. But availability from animal sources in larger quantities as well as the requirement for animal-free enzymes for pharmaceutical applications is a problem for such enzymes. Recently, a recombinant system for the production of pig liver esterase was described, but expression rates are still low (Lange *et al.*, 2001). A valuable alternative are microbial esterases. Different recombinant *E. coli* systems were developed to produce these enzymes.

In summary esterases are very interesting catalytic systems for industrial applications. These enzymes are characterized by high selectivity and activity combined with a broad substrate range. Hydrolysis reactions can be done at mild conditions. The features of the recombinant esterase of the psychrotolerant aerobic bacteria *Pseudoalteromonas arctica* can be summarized as follows: very low working temperature, relatively good thermostability in the neutral and alkaline range, a broad substrate range including chiral compounds, average stability, and activity in non-aqueous systems such as organic solvents.

5   SUMMARY

In this study, an aerobic psychrotolerant marine bacterium was isolated at 4 °C
from sea water samples collected from Spitzbergen in the Arctic. It was isolated
and characterized. This strain is a polarly flagellated Gram negative bacterium.
The strain grows optimally at a temperature of 10-15 °C and a pH range of 7-8
in media containing 2 to 3 % NaCl (w/v), using various carbohydrate and
organic acids as substrates. The fatty acid methyl ester (FAME) is composed of:
18.08 % straight chain saturated FAME, 9.61 % terminally branched saturated
FAME and 71.45% monounsaturated FAME. 16:0 straight chain saturated
FAME (8.23 %) and 16:1 w7c monounsaturated FAME (58.60 %) were the
most abundant FAMEs found. Phylogenetic analysis revealed a close
relationship of the new isolate to *Pseudoalteromonas elyakovii* with 99 % 16S rDNA
sequence identity, 47.6 % DNA-DNA similarity and G+C value of 39 % mol.
Phylogenetic evidence, together with phenotypic characteristics, show that this
isolate constitutes a novel species of the genus *Pseudoalteromonas* and the name
Pseudoalteromonas arctica is proposed.

A gene encoding an esterase (Est 37) was identified and sequenced from a gene
library of the psychrotolerant bacterium *Pseudoalteromonas arctica* sp. nov. The gene
1203 kb long, encodes an enzyme with molecular mass of 45 kDa (400 amino
acids). The amino acid sequence indicated that the enzyme has no signal peptide.
A hypothetical Shine-Dalgarno-like sequence with a weak motif (AAGG) was
found at 7 to 10 bp upstream from the predicted translational start codon. The
highly conserved five-residue motif typical of all α/β hydrolases (GxSxG) was
detected at position 104 to 108. A catalytic triad consisting of one serine, one
histidine and one aspartic acid residue was detected at positions 106, 225 and 196
respectively. It is therefore proposed that Est 37 belongs to the serine hydrolase
family. The ORF of the esterase gene has two domains; an esterase and an

OsmC like domain. The esterase gene exhibits 90 % amino acid identity with a hypothetical protein from *Pseudoalteromonas haloplanktis* containing similar esterase and OsmC domains. Est 37 sequence has less than 10% identity to the amino acid sequences of known esterases. It does not belong to any of the known α/β hydrolase families as well. The gene was cloned in pET-24b+ in frame with a His.Tag sequence and was actively expressed in *E. coli* Tuner™ (DE3) cells. The recombinant enzyme was purified to a final specific activity of 2.0 U/mg with a yield of 96%. Est 37 is active at a range of temperature of (0 – 30 °C) and pH of (6 – 9), showing optimum activity at 25 °C and pH 7.5-8. Est 37 retains 50% of its maximal activity at 0°C, and requires a very low activation energy E_a=5.1 kcal/mol. The enzyme is stable at temperatures of 4 and 10 °C. It has a half life of 15 hours at 30 °C and 5 hours at 40°C. The enzyme is stable at alkaline rather than at acidic pHs. The recombinant esterase exhibited activity towards substrates with short chains, such as p-nitrophenyl esters with fatty acid chain shorter that C8. Est 37 showed to be regio-selective. The recombinant esterase was able to hydrolyze several ester bonds in molecules bearing chiral carbons.

SUMMARY

105

BIBLIOGRAPHY

Abbott, J. C. (2001). Update on antiviral agents, California. *J Health Syst Pharmacy* July/August, 6.

Aghajari, N., Feller, G., Gerday, C. & Haser, R. (1998). Structures of the psychrophilic Alteromonas haloplanctis alpha-amylase give insights into cold adaptation at a molecular level. *Structure* **6**, 1503-16.

Altschul, S. F., Gish, W., Miller, W., Myers, E. W. & Lipman, D. J. (1990). Basic local alignment search tool. *J Mol Biol* **215**, 403-10.

Alvarez-Macarie, E., Augier-Magro, V. & Baratti, J. (1999). Characterization of a thermostable esterase activity from the moderate thermophile Bacillus licheniformis. *Biosci Biotechnol Biochem* **63**, 1865-70.

Arpigny, J. L., Feller, G. & Gerday, C. (1993). Cloning, sequence and structural features of a lipase from the antarctic facultative psychrophile Psychrobacter immobilis B10. *Biochim Biophys Acta* **1171**, 331-3.

Arpigny, J. L. & Jaeger, K. E. (1999). Bacterial lipolytic enzymes: classification and properties. *Biochem J* **343 Pt 1**, 177-83.

Arpigny, J. L., Lamotte, J. & Gerday, C. (1997). Molecular adaptation to cold of an Antarctic bacterial lipase. *J Mol Catal B* **3**, 29-35.

Asther, M., Haon, M., Roussos, S., Record, E., Delattre, M., Meessen, L., Labat, M. & Asther, M. (2002). Feruloyl esterase from *Aspegillus niger* a comparison of the production in solid state and submerged fermentation. *Process Biochem* **38**, 685-691.

Azzolina, O., Collina, S. & Vercesi, D. (1995a). Stereoselective hydrolysis by esterase: a strategy for resolving 2-(R,R´-phenoxy)propionyl ester racemates. *Il Farmanco* **50**, 725-733.

Azzolina, O., Vercesi, D., Collina, S. & Ghislandi, V. (1995b). Chiral resolution of methyl 2-aryloxypropionates by biocatalytic stereospecific hydrolysis. *Il Farmanco* **50**, 221-226.

Baron, A., Rombouts, F., Drilleau, J. F. & Pilnik, W. (1980). Purification et properties de la pectinesterase produite par *Apergillus niger. Lebensm Wiss technol* **13**, 330-333.

Baumann, L., Baumann, P., Mandel, M. & Allen, R. D. (1972). Taxonomy of aerobic marine eubacteria. *J Bacteriol* **110**, 402-29.

Baumann, M., Hauer, H. & Bornscheuer, U. (2000). Rapid screening hydrolases for the enantioselective conversion of difficult-to-resolve substrates. *Tetrahedron Asymm* **11**, 4781-4790.

Bentahir, M., Feller, G., Aittaleb, M., Lamotte-Brasseur, J., Himri, T., Chessa, J. P. & Gerday, C. (2000). Structural, kinetic, and calorimetric characterization of the cold-active phosphoglycerate kinase from the antarctic Pseudomonas sp. TACII18. *J Biol Chem* **275**, 11147-53.

Bentley, G. N., Jones, A. K. & Agnew, A. (2003). Mapping and sequencing of acetylcholinesterase genes from the platyhelminth blood fluke Schistosoma. *Gene* **314**, 103-12.

Biely, P., Cote, G. L., Kremnicky, L., Greene, R. V., Dupont, C. & Kluepfel, D. (1996). Substrate specificity and mode of action of acetylxylan esterase from Streptomyces lividans. *FEBS Lett* **396**, 257-60.

Bligh, E. G. & Dyer, W. J. (1959). A rapid method of total lipid extraction and purification. *Can J Biochem Physiol* **37**, 911-7.

Bohm, C., Austin, W. F. & Trauner, D. (2003). Enzymatic desymmetrization of a centrosymmetric diacetate. *Tetrahedron Asymm* **14**, 71-74.

Bornscheuer, U. T. (2002). Microbial carboxyl esterases: classification, properties and application in biocatalysis. *FEMS Microbiol Rev* **26**, 73-81.

Boublik, Y., Saint-Aguet, P., Lougarre, A., Arnaud, M., Villatte, F., Estrada-Mondaca, S. & Fournier, D. (2002). Acetylcholinesterase engineering for detection of insecticide residues. *Protein Eng* **15**, 43-50.

Bozal, N., Tudela, E., Rossello-Mora, R., Lalucat, J. & Guinea, J. (1997). Pseudoalteromonas antarctica sp. nov., isolated from an Antarctic coastal environment. *Int J Syst Bacteriol* **47**, 345-51.

Bradford, M. M. (1976). A rapid and sensitive method for the quantitation of microgram quantities of protein utilizing the principle of protein-dye binding. *Anal Biochem* **72**, 248-54.

Brady, L., Brzozowski, A. M., Derewenda, Z. S., Dodson, E., Dodson, G., Tolley, S., Turkenburg, J. P., Christiansen, L., Huge-Jensen, B., Norskov, L. & Et Al. (1990). A serine protease triad forms the catalytic centre of a triacylglycerol lipase. *Nature* **343**, 767-70.

Breeuwer, P., Drocourt, J. L., Bunschoten, N., Zwietering, M. H., Rombouts, F. M. & Abee, T. (1995). Characterization of uptake and hydrolysis of fluorescein diacetate and carboxyfluorescein diacetate by intracellular esterases in Saccharomyces cerevisiae, which result in accumulation of fluorescent product. *Appl Environ Microbiol* **61**, 1614-9.

Brenner, S. (1988). The molecular evolution of genes and proteins: a tale of two serines. *Nature* **334**, 528-30.

Brown, M., Davies, I. M., Moffat, C. F., Redshaw, J. & Craft, J. A. (2004). Characterisation of choline esterases and their tissue and subcellular distribution in mussel (Mytilus edulis). *Mar Environ Res* **57**, 155-69.

Buchholz-Cleven, B. E. E., Rattunde, B. & Staub, K. L. (1997). Screening forgenetic diversity of anaerobic Fe(II)- oxidizing bacteria using DGGE and whole-cell hybridization. *Syst Appl Microbiol* **20**, 301-309.

Calero-Rueda, O., Plou, F. J., Ballesteros, A., Martinez, A. T. & Martinez, M. J. (2002). Production, isolation and characterization of a sterol esterase from Ophiostoma piceae. *Biochim Biophys Acta* **1599**, 28-35.

Campon, B. & Kilbanon, A. M. (1984). Preparative production of optically active esters and alcohols using esterase-catalyzed stereospecific transesterification in organic media. *J Am Chem Soc* **106**, 2687-2692.

Carriere, F., Withers-Martinez, C., Van Tilbeurgh, H., Roussel, A., Cambillau, C. & Verger, R. (1998). Structural basis for the substrate selectivity of pancreatic lipases and some related proteins. *Biochim Biophys Acta* **1376**, 417-32.

Cashion, P., Holder-Franklin, M. A., Mccully, J. & Franklin, M. (1977). A rapid method for the base ratio determination of bacterial DNA. *Anal Biochem* **81**, 461-6.

Cavicchioli, R., Siddiqui, K. S., Andrews, D. & Sowers, K. R. (2002). Low-temperature extremophiles and their applications. *Curr Opin Biotechnol* **13**, 253-61.

Chaabouni, M., Pulvin, S., Touraud, D. & Thomas, D. (1996). Enzymatic sysnthesis of gerniol esters in a solvent free system by lipases. *Biotechnol lett* **18**, 1083-1088.

Chapus, C., Rovery, M., Sarda, L. & Verger, R. (1988). Minireview on pancreatic lipase and colipase. *Biochemie* **70**, 1223-1234.

Chavkin, C. (2004). Neural pathways transducing pain, non-narcotic analgesics. *Phamacology* **402/512**, 153-154.

Cheeseman, J. D., Tocilj, A., Park, S., Schrag, J. D. & Kazlauskas, R. J. (2004). Structure of an aryl esterase from Pseudomonas fluorescens. *Acta Crystallogr D Biol Crystallogr* **60**, 1237-43.

Chen, C. S., Fujimoto, Y., Girdaukas, G. & Sih, C. J. (1982). Quantitative analyses of biochemical kinetic resolutions of enantiomers. *J Am Chem Soc* **104**, 7294-7299.

Choi, Y. & Lee, B. (2001). Culture conditions for the production of esterase from *Lactobacillus casei* CL 96. *Bioprocess Biosyst Eng* **24**, 59-63.

Choo, D. W., Kurihara, T., Suzuki, T., Soda, K. & Esaki, N. (1998). A cold-adapted lipase of an Alaskan psychrotroph, Pseudomonas sp. strain B11-1: gene cloning and enzyme purification and characterization. *Appl Environ Microbiol* **64**, 486-91.

Christakopoulos, P., Tzalas, B., Mamma, D., Stamatis, H., Liadakis, G., Tzia, C., Kekos, D., Kolisis, F. & Macris, B. (1998). Production of esterases from *Fusarium oxysporum* catalyzing transesterification reactions in organic solvents. *Process Biochem* **33**, 729-733.

Christner, B. C., Kvitko, B. H., 2nd & Reeve, J. N. (2003). Molecular identification of bacteria and Eukarya inhabiting an Antarctic cryoconite hole. *Extremophiles* **7**, 177-83.

Claudianos, C., Russell, R. J. & Oakeshott, J. G. (1999). The same amino acid substitution in orthologous esterases confers organophosphate resistance on the house fly and a blowfly. *Insect Biochem Mol Biol* **29**, 675-86.

Coppens, M., Versichelen, L. & Mortier, E. (2002). Treatment of postoperative pain after ophthalmic surgery. *Bull Soc Belge Ophtalmol* 27-32.

Costenoble, R., Adler, L., Niklasson, C. & Liden, G. (2003). Engineering of the metabolism of Saccharomyces cerevisiae for anaerobic production of mannitol. *FEMS Yeast Res* **3**, 17-25.

Cygler, M., Schrag, J. D., Sussman, J. L., Harel, M., Silman, I., Gentry, M. K. & Doctor, B. P. (1993). Relationship between sequence conservation and three dimentional structure in a large family of esterases, lipases, and related proteins. *Protein Sci.* **2**, 366-382.

D'amico, S., Claverie, P., Collins, T., Georlette, D., Gratia, E., Hoyoux, A., Meuwis, M. A., Feller, G. & Gerday, C. (2002). Molecular basis of cold adaptation. *Philos Trans R Soc Lond B Biol Sci* **357**, 917-25.

Davail, S., Feller, G., Narinx, E. & Gerday, C. (1994). Cold adaptation of proteins. Purification, characterization, and sequence of the heat-labile subtilisin from the antarctic psychrophile Bacillus TA41. *J Biol Chem* **269**, 17448-53.

De Ley, J., Cattoir, H. & Reynaerts, A. (1970). The quantitative measurement of DNA hybridization from renaturation rates. *Eur J Biochem* **12**, 133-42.

De Ruiter, C. J. & De Haan, A. (2003). Shortening-induced depression of voluntary force in unfatigued and fatigued human adductor pollicis muscle. *J Appl Physiol* **94**, 69-74.

De Vries, R. P. & Visser, J. (1999). Regulation of the feruloyl esterase (faeA) gene from Aspergillus niger. *Appl Environ Microbiol* **65**, 5500-3.

Deming, J. W. (2002). Psychrophiles and polar regions. *Curr Opin Microbiol* **5**, 301-9.

Drablos, F. & Petersen, S. B. (1997). Identification of conserved residues in family of esterase and lipase sequences. *Methods Enzymol* **284**, 28-61.

Dufour, J. & Bing, Y. (2001). Influence of yeast strain and fermentation conditions on yeast esterase activities. *Brew Dig* **76**, 44.

Duphar-Int-Res (1994). Stereospecific hetero-bicyclic alcohol enantiomer preparation. **European Patent**, 605033.

Ewis, H. E., Abdelal, A. T. & Lu, C. D. (2004). Molecular cloning and characterization of two thermostable carboxyl esterases from Geobacillus stearothermophilus. *Gene* **329**, 187-95.

Faulds, C. B., Devries, R. P., Kroon, P. A., Visser, J. & Williamson, G. (1997). Influence of ferulic acid on the production of feruloyl esterases by Aspergillus niger. *FEMS Microbiol Lett* **157**, 239-44.

Feller, G. (2003). Molecular adaptations to cold in psychrophilic enzymes. *Cell Mol Life Sci* **60**, 648-62.

Feller, G. & Gerday, C. (1997). Psychrophilic enzymes: molecular basis of cold adaptation. *Cell Mol Life Sci* **53**, 830-41.

Feller, G. & Gerday, C. (2003). Psychrophilic enzymes: hot topics in cold adaptation. *Nat Rev Microbiol* **1**, 200-8.

Feller, G., Le Bussy, O. & Gerday, C. (1998). Expression of psychrophilic genes in mesophilic hosts: assessment of the folding state of a recombinant alpha-amylase. *Appl Environ Microbiol* **64**, 1163-5.

Feller, G., Lonhienne, T., Deroanne, C., Libioulle, C., Van Beeumen, J. & Gerday, C. (1992). Purification, characterization, and nucleotide sequence of the thermolabile alpha-amylase from the antarctic psychrotroph Alteromonas haloplanctis A23. *J Biol Chem* **267**, 5217-21.

Feller, G., Narinx, E., Aprigny, J. L., Aittaleb, M., Baise, E., Genicot, S. & Gerday, C. (1989). Enzymes from psychrophilic organisms. *FEMS Microbiol Rev* **18**, 189-202.

Feller, G., Payan, F., Theys, F., Qian, M., Haser, R. & Gerday, C. (1994). Stability and structural analysis of alpha-amylase from the antarctic psychrophile Alteromonas haloplanctis A23. *Eur J Biochem* **222**, 441-7.

Felsenstien, J. (1995). PHYLIP (Phylogeny Inference Package), version 3.65. *Edited by Department of Genetics, University of Washington, Seattle, WA, USA*

Fenster, K. M., Parkin, K. L. & Steels, J. L. (2003). Intracellular esterase from Lactobacillus casei LILA: nucleotide sequencing, purification, and characterization. *J Dairy Sci* **86**, 1118-29.

Fernandez, J., Mohedano, A. F., Fernandez-Garcia, E., Medina, M. & Nunez, M. (2004). Purification and characterization of an extracellular tributyrin esterase produced by a cheese isolate, *Micrococcus* sp. INIA 528. *Int. Dairy J* **14**, 135-142.

Ferrer, M., Chernikova, T. N., Timmis, K. N. & Golyshin, P. N. (2004). Expression of a temperature-sensitive esterase in a novel chaperone-based Escherichia coli strain. *Appl Environ Microbiol* **70**, 4499-504.

Fillingham, I. J., Kroon, P. A., Williamson, G., Gilbert, H. J. & Hazlewood, G. P. (1999). A modular cinnamoyl ester hydrolase from the anaerobic fungus Piromyces equi acts synergistically with xylanase and is part of a multiprotein cellulose-binding cellulase-hemicellulase complex. *Biochem J* **343 Pt 1**, 215-24.

Finer, Y., Jaffer, F. & Santerre, J. P. (2004). Mutual influence of cholesterol esterase and pseudocholinesterase on the biodegradation of dental composites. *Biomaterials* **25**, 1787-93.

Foght, J., Aislabie, J., Turner, S., Brown, C. E., Ryburn, J., Saul, D. J. & Lawson, W. (2004). Culturable Bacteria in Subglacial Sediments and Ice from Two Southern Hemisphere Glaciers. *Springer-Verlag: New York;* **47**, 329-340.

Forster, J. (1887). Ueber einige Eingeschaften leuchtender Bakterien. *Centr Bakteriol Parasitenk* **2**, 337-340.

Fu, P. P., Howard, P. C., Culp, S. J., Xia, Q., Webb, P. J., Blankenship, L. R., Wamer, W. G. & Bucher, J. R. (2002). Do topically applied skin creams containing retinyl palmitate affect the photocarcinogenecity of simulated solar light. *J Food Drug Anal* **10**, 262-268.

Fukuda, K., Yamamoto, N., Kiyokawa, Y., Yanagiuchi, T., Wakai, Y., Kitamoto, K., Inoue, Y. & Kimura, A. (1998). Balance of activities of alcohol acetyltransferase and esterase in Saccharomyces cerevisiae is important for production of isoamyl acetate. *Appl Environ Microbiol* **64**, 4076-8.

Gao, R., Y, F., K, I., H, I., S, A., Y, K. & S, C. (2003). Cloning, purification and properties of a hyperthermophilic esterase from archaeon *Aerophyrum pernix* KL. *J Mol Cata* **B 24-25**, 1-8.

Garcia, B. L., Ball, A. S., Rodriguez, J., Perez-Leblic, M. I., Arias, M. E. & Copa-Patino, J. L. (1998). Induction of ferulic acid esterase and xylanase activities in *Streptomyces avermitilis* UAH 30. *FEMS Microbiol Lett* **158**, 95-99.

Gauthier, G., Gauthier, M. & Christen, R. (1995). Phylogenetic analysis of the genera Alteromonas, Shewanella, and Moritella using genes coding for small-subunit rRNA sequences and division of the genus Alteromonas into two genera, Alteromonas (emended) and Pseudoalteromonas gen. nov., and proposal of twelve new species combinations. *Int J Syst Bacteriol* **45**, 755-61.

Gerday, C., Aittaleb, M., Bentahir, M., Chessa, J. P., Claverie, P., Collins, T., D'amico, S., Dumont, J., Garsoux, G., Georlette, D., Hoyoux, A., Lonhienne, T., Meuwis, M. A. & Feller, G. (2000). Cold-adapted enzymes: from fundamentals to biotechnology. *Trends Biotechnol* **18**, 103-7.

Gerhardt, P., Murray, R. G. E., Wood, W. A. & Krieg, N. R. (1994). Method of General and Molecular Microbiology. *American Society for Microbiology* Washington DC.

Gianese, G., Argos, P. & Pascarella, S. (2001). Structural adaptation of enzymes to low temperatures. *Protein Eng* **14**, 141-8.

Giuliani, S., Piana, C., Setti, L., Hochkoeppler, A., Pifferi, P., Williamson, G. & Faulds, C. (2001). Synthesis of pentylferulate by a feruloyl esterase from *Aspergillus niger* using water-in-oil microemulsions. *Biotechnol Lett* **23**, 325-330.

Gokul, B. (1999). Studies on suitable organism expressing esterase specific for enzymatic resolution of (R,S)-Beta-acetylmercaptoisobutyrate. *IIT Madras* **Ph.D.**,

Gudelj, M., Valinger, G., Faber, K. & Schwab, H. (1998). Novel *Rhodococcus* esterases by genetic engineering. *J Mol Cata B* **5**, 261-266.

Gutierrez, C. & Devedjian, J. C. (1991). Osmotic induction of gene osmC expression in Escherichia coli K12. *J Mol Biol* **220**, 959-73.

He, Y. P., Ma, E. B. & Zhu, K. Y. (2004). Characterization of general esterases in relation to malation susceptibility in two field populations of the oriental migratory locust, *locusta migratoria manilensis* (Meyen). *Pesticide Biochem Physiol* **78**, 103-113.

Heidari, R., Devonshire, A. L., Campbell, B. E., Bell, K. L., Dorrian, S. J., Oakeshott, J. G. & Russell, R. J. (2004). Hydrolysis of organophosphorus insecticides by in vitro modified carboxylesterase E3 from Lucilia cuprina. *Insect Biochem Mol Biol* **34**, 353-63.

Henke, E. & Bornscheuer, U. T. (2002). Esterases from Bacillus subtilis and B. stearothermophilus share high sequence homology but differ substantially in their properties. *Appl Microbiol Biotechnol* **60**, 320-6.

Hofmann, K., Bucher, P., Falquet, L. & Bairoch, A. (1999). The PROSITE database, its status in 1999. *Nucleic Acids Res* **27**, 215-9.

Holmquist, M. (2000). Alpha/Beta-hydrolase fold enzymes: structures, functions and mechanisms. *Curr Protein Pept Sci* **1**, 209-35.

Horne, I., Harcourt, R. L., Sutherland, T. D., Russell, R. J. & Oakeshott, J. G. (2002). Isolation of a Pseudomonas monteilli strain with a novel phosphotriesterase. *FEMS Microbiol Lett* **206**, 51-5.

Horsman, G. P., Liu, A. M., Henke, E., Bornscheuer, U. T. & Kazlauskas, R. J. (2003). Mutations in distant residues moderately increase the enantioselectivity of Pseudomonas fluorescens esterase towards methyl 3bromo-2-methylpropanoate and ethyl 3phenylbutyrate. *Chemistry* **9**, 1933-9.

Howard, G., Crother, B. & Vicknair, J. (2001). Cloning, nucleotide sequencing and characterization of a polyurethanase gene (*pue*B) from *Pseudomonas chlororaphis. int Biodeterior Biodegrad* **47**, 141-149.

Huss, V. A. R., Festl, H. & Schleifer, K. H. (1983). Studies on the spectrophotometric determination of DNA hybridization from renaturation rates. *Syst Appl Microbiol* **4**, 184-192.

Ivanova, E. P., Bakunina, I. Y., Nedashkovskaya, O. I., Gorshkova, N. M., Alexeeva, Y. V., Zelepuga, E. A., Zvaygintseva, T. N., Nicolau, D. V. & Mikhailov, V. V. (2003). Ecophysiological variabilities in ectohydrolytic enzyme activities of some Pseudoalteromonas species, P. citrea, P. issachenkonii, and P. nigrifaciens. *Curr Microbiol* **46**, 6-10.

Ivanova, E. P., Chun, J., Romanenko, L. A., Matte, M. E., Mikhailov, V. V., Frolova, G. M., Huq, A. & Colwell, R. R. (2000). Reclassification of Alteromonas distincta Romanenko et al. 1995 as Pseudoalteromonas distincta comb. nov. *Int J Syst Evol Microbiol* **50 Pt 1**, 141-4.

Ivanova, E. P., Gorshkova, N. M., Zhukova, N. V., Lysenko, A. M., Zelepuga, E. A., Prokof'eva, N. G., Mikhailov, V. V., Nicolau, D. V. & Christen, R. (2004). Characterization of Pseudoalteromonas distincta-like seawater isolates and description of Pseudoalteromonas aliena sp. nov. *Int J Syst Evol Microbiol* **54**, 1431-7.

Ivanova, E. P., Kiprianova, E. A., Mikhailov, V. V., Levanova, G. F., Garagulya, A. D., Gorshkova, N. M., Vysotskii, M. V., Nicolau, D. V., Yumoto, N., Taguchi, T. & Yoshikawa, S. (1998). Phenotypic diversity of Pseudoalteromonas citrea from different marine habitats and emendation of the description. *Int J Syst Bacteriol* **48 Pt 1**, 247-56.

Ivanova, E. P., Mikhailov, V. V., Kiprianova, E. A., Levanova, G. F., Garagulya, A. D., Frolova, G. M. & Svetashev, V. I. (1996). *Alteromonas elyakovii* sp. nov. a new bacterium isolated from marine mollusks. *Russ J Mar Biol* **22**, 209-215.

Jaeger, K. E., Dijkstra, B. W. & Reetz, M. T. (1999). Bacterial biocatalysts: molecular biology, three-dimensional structures, and biotechnological applications of lipases. *Annu Rev Microbiol* **53**, 315-51.

Jaeger, K. E., Eggert, T., Eipper, A. & Reetz, M. T. (2001). Directed evolution and the creation of enantioselective biocatalysts. *Appl Microbiol Biotechnol* **55**, 519-30.

Jaeger, K. E., Ransac, S., Dijkstra, B. W., Colson, C., Van Heuvel, M. & Misset, O. (1994). Bacterial lipases. *FEMS Microbiol Rev* **15**, 29-63.

Jahangir, R., Mccloskey, C. B., Mc Clung, W. G., Labow, R. S., Brash, J. L. & Santerre, J. P. (2003). The influence of protein adsorption and surface modifying macromolecules on the hydrolytic degradation of a poly(ether-urethane) by cholesterol esterase. *Biomaterials* **24**, 121-30.

Jahnke, K. D. (1992). Basic computer for evaluation of spectoscopic DNA renaturation data from GILFORD System 2600 spectrometer on a PC/XT/AT type personal computer. *J Micobiol Methods* **15**, 61-73.

John, G. & Heinzle, E. (2000). Quantitative screening method for hydrolysis in microplates using pH indicators: determination of kinetic parameters by dynamic monitoring. *Biotechnol Bioeng* **72**, 620-627.

Juge, N., Williamson, G., Puigserver, A., Cummings, N. J., Connerton, I. F. & Faulds, C. B. (2001). High-level production of recombinant Aspergillus niger cinnamoyl esterase (FAEA) in the methylotrophic yeast Pichia pastoris. *FEMS Yeast Res* **1**, 127-32.

Jung, Y. J., Lee, J. K., Sung, C. G., Oh, T. K. & Kim, H. K. (2003). nonionic detergent-induced activation of an esterase from *Bacillus megaterium* 20-1. *J Mol Cata* **26**, 223-229.

Junge, W. & Krisch, K. (1973). Current problems on the structure and classification of mammalian liver carboxylesterases (EC 3.1.1.1). *Mol Cell Biochem* **1**, 41-52.

Kademi, A., Ait-Abdelkader, N., Fakhreddine, L. & Baratti, J. (2000). Purification and characterization of a thermostable esterase from the moderate thermophile Bacillus circulans. *Appl Microbiol Biotechnol* **54**, 173-9.

Kanegafuchi-Chem (1991). New stereospecific 5-acyloxymethyl-oxazolidin-2-one derivative. **Patent**, J03108498.

Kawanami, T., Miyakoshi, M., Dairi, T. & Itoh, N. (2002). Reaction mechanism of the Co2+-activated multifunctional bromoperoxidase-esterase from Pseudomonas putida IF-3. *Arch Biochem Biophys* **398**, 94-100.

Kermasha, S., Bisakowski, B., Ismail, S. & Morin, A. (2000). The effect of physical and chemical treatments on the esterase activity from *Pseudomonas fragi* CRDA 037. *Food Res Int* **33**, 767-774.

Kim, G. J., Choi, G. S., Kim, J. Y., Lee, J. B., Jo, D. H. & Ryu, Y. W. (2002a). Screening production and properties of a stereo specific esterase from *pseudomonas* species-34 with high selectivity to (S)-ketoprofen ethyl ester. *J Mol. Catal B* **17**, 29-38.

Kim, H., Na, H., Park, M., Oh, T. & Lee, T. (2004). Occurence of ofloxacin ester hydrolyzing esterase from *Bacillus niacini* EM001. *J Mol Catal B* **27**, 237-241.

Kim, K. K., Song, H. K., Shin, D. H., Hwang, K. Y., Choe, S., Yoo, O. J. & Suh, S. W. (1997). Crystal structure of carboxylesterase from Pseudomonas fluorescens, an alpha/beta hydrolase with broad substrate specificity. *Structure* **5**, 1571-84.

Kim, Y. H., Cha, C. J. & Cerniglia, C. E. (2002b). Purification and characterization of an erythromycin esterase from an erythromycin-resistant Pseudomonas sp. *FEMS Microbiol Lett* **210**, 239-44.

Kimura, M. (1980). A simple method for estimating evolutionary rates of base substitutions through comparative studies of nucleotide sequences. *J Mol Evol* **16**, 111-20.

Kohn, G., Van Der Ploeg, P., Mobius, M. & Sawatzki, G. (1996). Influence of the derivatization procedure on the results of the gaschromatographic fatty acid analysis of human milk and infant formulae. *Z Ernahrungswiss* **35**, 226-34.

Kontkanen, H., Tenkanen, M., Fagerstrom, R. & Reinikainen, T. (2004). Characterisation of steryl esterase activities in commercial lipase preparations. *J Biotechnol* **108**, 51-9.

Krebsfanger, N., Schierholz, K. & Bornscheuer, U. T. (1998). Enantioselectivity of a recombinant esterase from Pseudomonas fluorescens towards alcohols and carboxylic acids. *J Biotechnol* **60**, 105-11.

Kroon, P. A., Faulds, C. B. & Williamson, G. (1996). Purification and characterization of a novel esterase induced by growth of Aspergillus niger on sugar-beet pulp. *Biotechnol Appl Biochem* **23** (Pt 3), 255-62.

Kulakova, L., Galkin, A., Nakayama, T., Nishino, T. & Esaki, N. (2004). Cold-active esterase from Psychrobacter sp. Ant300: gene cloning, characterization, and the effects of Gly-->Pro substitution near the active site on its catalytic activity and stability. *Biochim Biophys Acta* **1696**, 59-65.

Laemmli, U. K. (1970). Cleavage of structural proteins during the assembly of the head of bacteriophage T4. *Nature* **227**, 680-5.

Lange, S., Musidlowska, A., Schmidt-Dannert, C., Schmitt, J. & Bornscheuer, U. T. (2001). Cloning, functional expression, and characterization of recombinant pig liver esterase. *Chembiochem* **2**, 576-82.

Laranja, A., Manzatto, A. & Bicudo, H. D. C. (2003). effects of caffeine and used coffee grounds on biological features of *Aedes aegypti* (Diptera, Culicidae) and their possible use in alternative control. *Genetics Mol Biol* **26**, 419-429.

Lepage, G. & Roy, C. C. (1984). Improved recovery of fatty acid through direct transesterification without prior extraction or purification. *J Lipid Res* **25**, 1391-6.

Levasseur, A., Benoit, I., Asther, M., Asther, M. & Record, E. (2004). Homologous expression of the feruloyl esterase B gene from Aspergillus niger and characterization of the recombinant enzyme. *Protein Expr Purif* **37**, 126-33.

Liu, A., Somers, N., Kazlauskas, R., Brush, T., Zocher, F., Enzelberger, M., Bornscheuer, U., Horsman, G., Mezzetti, A., Schmidt-Dannert, C. & Schmid, R. (2001). Mapping the substrate selectivity of new hydrolases using colorimetric screening: lipases from *Bacillus thermocatenulatus* and *Ophiostoma piliferum*, esterases from *Pseudomonas fluorescens* and *Stryptomyces diastochromogenes*. *Tetrahedron Asymm* **12**, 545-556.

Lomolino, G., Rizzi, C., Spettoli, P., Curioni, A. & Lante, A. (2003). Cell vitality and esterase activity of *Saccharomyces cerevisiae* is affected by increasing calcium concentration. *Biotechnology* **(Nov/Dec)**, 32-35.

Lonhienne, T., Gerday, C. & Feller, G. (2000). Psychrophilic enzymes: revisiting the thermodynamic parameters of activation may explain local flexibility. *Biochim Biophys Acta* **1543**, 1-10.

Ma, J., Campbell, A. & Karlin, S. (2002). Correlations between Shine-Dalgarno sequences and gene features such as predicted expression levels and operon structures. *J Bacteriol* **184**, 5733-45.

Margesin, R. & Shinner, F. (1994). Properties of cold-adapted microorganisms and their potential role in biotechnology. *J Biotech* **33**, 1-14.

Margesin, R. F., G. Gerday,C. Russel, N.J. (2002). Cold adapted microorganisms: adaptation strategies and biotechnological potetial. *The encyclopedia of environmental microbiology* 871-885.

Mesbah, M. & Whitman, W. B. (1989). Measurement of deoxyguanosine/thymidine ratios in complex mixtures by high-performance liquid chromatography for determination of the mole percentage guanine + cytosine of DNA. *J Chromatogr* **479**, 297-306.

Miteva, V. I., Sheridan, P. P. & Brenchley, J. E. (2004). Phylogenetic and physiological diversity of microorganisms isolated from a deep greenland glacier ice core. *Appl Environ Microbiol* **70**, 202-13.

Molinavi, F., Brenna, O., Valenti, M. & Aragozzini, F. (1996). Isolation of a novel carboxylesterase from *Bacillus coagulans* with high enantioselectivity toward racemic esters of 1,2-O-isopropylideneglycerol. *Enzyme Microbiol Technol* **19**, 551-556.

Moore, J. C. & Arnold, F. H. (1996). Directed evolution of a para-nitrobenzyl esterase for aqueous-organic solvents. *Nat Biotechnol* **14**, 458-67.

Morana, A., Di Prizito, N., Aurilia, V., Rossi, M. & Cannio, R. (2002). A carboxylesterase from the hyperthermophilic archaeon Sulfolobus solfataricus: cloning of the gene, characterization of the protein. *Gene* **283**, 107-15.

Morita, R. Y. (1975). Psychrophilic bacteria. *Bacteriol Rev* **39**, 144-167.

Musidlowska-Persson, A. & Bornscheuer, U. T. (2003). Recombinant porcine intestinal carboxylesterase: cloning from the pig liver esterase gene by site-directed mutagenesis, functional expression and characterization. *Protein Eng* **16**, 1139-45.

Mustranta, A. (1992). Use of lipases in the resolution of racemic ibuprofen. *Appl Microbiol Biotechnol* **38**, 61-6.

Napolitano, M. J. & Shain, D. H. (2004). Four kingdoms on glacier ice: convergent energetic processes boost energy levels as temperatures fall. *Proc Biol Sci* **271 Suppl 5**, S273-6.

Nardini, M. & Dijkstra, B. W. (1999). Alpha/beta hydrolase fold enzymes: the family keeps growing. *Curr Opin Struct Biol* **9**, 732-7.

Narinx, E., Baise, E. & Gerday, C. (1997). Subtilisin from psychrophilic antarctic bacteria: characterization and site-directed mutagenesis of residues possibly involved in the adaptation to cold. *Protein Eng* **10**, 1271-9.

Nishizawa, M., Gomi, H. & Kishimoto, F. (1993). Purification and some properties of carboxylesterase from *Arthobacter globiformis;* stereoselective hydrolysis of ethyl chrysanthemate. *Biosci Biotechnol Biochem* **57**, 594-598.

Nishizawa, M., Shimizu, M., Ohkawa, H. & Kanaoka, M. (1995). Stereoselective production of (+)-trans-chrysanthemic acid by a microbial esterase: cloning, nucleotide sequence, and overexpression of the esterase gene of Arthrobacter globiformis in Escherichia coli. *Appl Environ Microbiol* **61**, 3208-15.

Oakeshott, J. G., Claudianos, C., Russell, R. J. & Robin, G. C. (1999). Carboxyl/cholinesterases: a case study of the evolution of a successful multigene family. *Bioessays* **21**, 1031-42.

Ollis, D. L., Cheah, E., Cygler, M., Dijkstra, B., Frolow, F., Franken, S. M., Harel, M., Remington, S. J., Silman, I., Schrag, J. & Et Al. (1992). The alpha/beta hydrolase fold. *Protein Eng* **5**, 197-211.

Ostdal, H., Baron, C., Blom, H. & Andersen, H. (1996). Prpduction isolation and partial characterization of lipase-esterase from *pediococcus pentosaceus* SV61. *Lebensm Wiss technol* **29**, 542-546.

Ozaki, E. & Sakashita, K. (1997). Esterase catalyzed region-and enantioselective hydrolysis of substituted carboxylases. *Chem Lett* **8**, 741-742.

Ozaki, E., Sakimae, A. & Numazawa, R. (1994). Cloning and expression of Pseudomonas putida esterase gene in Escherichia coli and its use in enzymatic production of D-beta-acetylthioisobutyric acid. *Biosci Biotechnol Biochem* **58**, 1745-6.

Panda, S. & Sahu, S. K. (2004). Recovery of acetylcholine esterase activity of Drawida willsi (Oligochaeta) following application of three pesticides to soil. *Chemosphere* **55**, 283-90.

Panda, T. & Gowrishankar, B. S. (2005). Production and applications of esterases. *Appl Microbiol Biotechnol* **67**, 160-9.

Patel, R. N. (2000). Stereoselective biocatalysis. Edited by Dekker, Newyork

Petersen, E. I., Valinger, G., Solkner, B., Stubenrauch, G. & Schwab, H. (2001). A novel esterase from Burkholderia gladioli which shows high deacetylation activity on cephalosporins is related to beta-lactamases and DD-peptidases. *J Biotechnol* **89**, 11-25.

Peterson, E., Valinger, G., Solkner, B., Stubenrauch, G. & Schwab, H. (2001). A novel esterase from *Bulkholderia gladioli* which shows high deacetylation activity on Cephalosporins is related to Beta-lactamases and DD-peptidases. *J Biotechnol* **89**, 11-25.

Picollo, M. I., Vassena, C. V., Mougabure Cueto, G. A., Vernetti, M. & Zerba, E. N. (2000). Resistance to insecticides and effect of synergists on permethrin toxicity in Pediculus capitis (Anoplura: Pediculidae) from Buenos Aires. *J Med Entomol* **37**, 721-5.

Pleiss, J., Fischer, M. & Schmid, R. D. (1998). Anatomy of lipase binding sites: the scissile fatty acid binding site. *Chem Phys Lipids* **93**, 67-80.

Pohlenz, H. D., Boidol, W., Schuttke, I. & Streber, W. R. (1992). Purification and properties of an Arthrobacter oxydans P52 carbamate hydrolase specific for the herbicide phenmedipham and nucleotide sequence of the corresponding gene. *J Bacteriol* **174**, 6600-7.

Poutanen, K., Sundberg, M., Korte, H. & Puls, J. (1990). Deacylation of xylans by acetyl esterases of *Trichoderma reesei*. *Appl Microbiol Biotechnol* **33**, 506-510.

Priefert, H., Rabenhorst, J. & Steinbuchel, A. (1997). Molecular characterization of genes of Pseudomonas sp. strain HR199 involved in bioconversion of vanillin to protocatechuate. *J Bacteriol* **179**, 2595-607.

Priscu, J. C., Adams, E. E., Lyons, W. B., Voytek, M. A., Mogk, D. W., Brown, R. L., Mckay, C. P., Takacs, C. D., Welch, K. A., Wolf, C. F., Kirshtein, J. D. & Avci, R. (1999). Geomicrobiology of subglacial ice above Lake Vostok, Antarctica. *Science* **286**, 2141-4.

Puls, J., Altaner, C. & Saake, B. (2001). *8th International Conference on Biotechnology in the Pulp and Paper Industry* 27-28.

Quax, W. J. & Broekhuizen, C. P. (1994). Development of a new Bacillus carboxyl esterase for use in the resolution of chiral drugs. *Appl Microbiol Biotechnol* **41**, 425-31.

Rainey, F. A. & Stackebrandt, E. (1993). 16S rDNA analysis reveals phylogenetic diversity among the polysaccharolytic clostridia. *FEMS Microbiol Lett* **113**, 125-8.

Reetz, M. (2000). Application of directed evolution in the development of enantioselective enzymes. *Pure Appl. Chem.* **72**, 1615-1622.

Reetz, M. T. & Jaeger, K. E. (1998). Overexpression, immobilization and biotechnological application of Pseudomonas lipases. *Chem Phys Lipids* **93**, 3-14.

Rehse, P. H., Ohshima, N., Nodake, Y. & Tahirov, T. H. (2004). Crystallographic structure and biochemical analysis of the Thermus thermophilus osmotically inducible protein C. *J Mol Biol* **338**, 959-68.

Rentier-Delrue, F., Mande, S. C., Moyens, S., Terpstra, P., Mainfroid, V., Goraj, K., Lion, M., Hol, W. G. & Martial, J. A. (1993). Cloning and overexpression of the triosephosphate isomerase genes from psychrophilic and thermophilic bacteria. Structural comparison of the predicted protein sequences. *J Mol Biol* **229**, 85-93.

Riegels, M., Koch, R., Pedersen, L. & Lund, H. (1997). Enzymatic hydrolysis of cyclic oligomers of poly(ethylene terephthalate). *WO* 9727237.

Romanenko, L. A., Zhukova, N. V., Rohde, M., Lysenko, A. M., Mikhailov, V. V. & Stackebrandt, E. (2003). Pseudoalteromonas agarivorans sp. nov., a novel marine agarolytic bacterium. *Int J Syst Evol Microbiol* **53**, 125-31.

Russell, N. J. (1990). Cold adaptation of microorganisms. *Philos Trans R Soc Lond B Biol Sci* **326**, 595-608, discussion 608-11.

Russell, R. J., Gerike, U., Danson, M. J., Hough, D. W. & Taylor, G. L. (1998). Structural adaptations of the cold-active citrate synthase from an Antarctic bacterium. *Structure* **6**, 351-61.

Satoh, T., Taylor, P., Bosron, W. F., Sanghani, S. P., Hosokawa, M. & La Du, B. N. (2002). Current progress on esterases: from molecular structure to function. *Drug Metab Dispos* **30**, 488-93.

Sawabe, T., Tanaka, R., Iqbal, M. M., Tajima, K., Ezura, Y., Ivanova, E. P. & Christen, R. (2000). Assignment of Alteromonas elyakovii KMM 162T and five strains isolated from spot-wounded fronds of Laminaria japonica to Pseudoalteromonas elyakovii comb. nov. and the extended description of the species. *Int J Syst Evol Microbiol* **50 Pt 1**, 265-71.

Schmid, R. D. & Verger, R. (1998). Lipases: interfacial enzymes with attractive applications. 1608-1633.

Schmidt-Dannert, C., Sztajer, H., Stocklein, W., Menge, U. & Schmid, R. D. (1994). Screening, purification and properties of a thermophilic lipase from Bacillus thermocatenulatus. *Biochim Biophys Acta* **1214**, 43-53.

Schmidt-Nielsen, S. (1902). *Centrbl Bakteriol Parasitenkd II Abt* **9**,

Schrag, J. D. & Cygler, M. (1997). Lipases and alpha/beta hydrolase fold. *Methods Enzymol* **284**, 85-107.

Schutt, H. (1996). Enantiomeric chroman carboxylic acid derivative production. DE, 4430089.

Shain, D. H., Mason, T. A., Farrell, A. H. & Michalewicz, L. A. (2001). Distribution and behavior of ice worms (Mesenchytraeus solifungus) in south-central Alaska. *Canadian Journal of Zoology* **79**, 1813-1821.

Shankaranand, V., Ramesh, M. & Lonsane, B. (1992). Idiosyncrasies of solid-state fermentation systems in the biosynthesis of metabolites by some baterial and fungal cultures. *Process Biochem* **27**, 33-36.

Shaw, E., Mccue, L. A., Lawrence, C. E. & Dordick, J. S. (2002). Identification of a novel class in the alpha/beta hydrolase fold superfamily: the N-myc differentiation-related proteins. *Proteins* **47**, 163-8.

Shen, D., Xu, J. H., Wu, H. Y. & Liu, Y. Y. (2002). Significantly improved esterase activity of Trichosporon brassicae cells for ketoprofen resolution by 2-propanol treatment. *J Mol Catal B* **18**, 219-224.

Simonen, M. & Palva, I. (1993). Protein secretion in Bacillus species. *Microbiol Rev* **57**, 109-37.

Skidmore, M., Anderson, S. P., Sharp, M., Foght, J. & Lanoil, B. D. (2005). Comparison of microbial community compositions of two subglacial environments reveals a possible role for microbes in chemical weathering processes. *Appl Environ Microbiol* **71**, 6986-97.

Skidmore, M. L., Foght, J. M. & Sharp, M. J. (2000). Microbial life beneath a high arctic glacier. *Appl Environ Microbiol* **66**, 3214-20.

Squibb (1993). Enzymatic resolution of taxane side chain intermediate. European Patent, 552041.

Staley, J. T. & Gosink, J. J. (1999). Poles apart: biodiversity and biogeography of sea ice bacteria. *Annu Rev Microbiol* **53**, 189-215.

Stuhlfelder, C., Lottspeich, F. & Mueller, M. J. (2002). Purification and partial amino acid sequences of an esterase from tomato. *Phytochemistry* **60**, 233-40.

Suzuki, T., Nakayama, T., Choo, D. W., Hirano, Y., Kurihara, T., Nishino, T. & Esaki, N. (2003). Cloning, heterologous expression, renaturation, and characterization of a cold-adapted esterase with unique primary structure from a psychrotroph Pseudomonas sp. strain B11-1. *Protein Expr Purif* **30**, 171-8.

Suzuki, T., Nakayama, T., Kurihara, T., Nishino, T. & Esaki, N. (2001). Cold-active lipolytic activity of psychrotrophic Acinetobacter sp. strain no. 6. *J Biosci Bioeng* **92**, 144-8.

Suzuki, T., Nakayama, T., Kurihara, T., Nishino, T. & Esaki, N. (2002). Primary structure and catalytic properties of a cold-active esterase from a psychrotroph, Acinetobacter sp. strain No. 6. isolated from Siberian soil. *Biosci Biotechnol Biochem* **66**, 1682-90.

Suzuki, Y., Miyamoto, K. & Ohta, H. (2004). A novel thermostable esterase from the thermoacidophilic archaeon Sulfolobus tokodaii strain 7. *FEMS Microbiol Lett* **236**, 97-102.

Tang, Y. W., Santerre, J. P., Labow, R. S. & Taylor, D. G. (1997). Application of macromolecular additives to reduce the hydrolytic degradation of polyurethanes by lysosomal enzymes. *Biomaterials* **18**, 37-45.

Tenkanen, M., Thornton, J. & Viikari, L. (1995). An acetylglucomannan esterase of Aspergillus oryzae; purification, characterization and role in the hydrolysis of O-acetyl-galactoglucomannan. *J Biotechnol* **42**, 197-206.

Tomioka, H. (1983). Purification and characterization of the tween-hydrolyzing esterase of Mycobacterium smegmatis. *J Bacteriol* **155**, 1249-59.

Topakas, E., Kalogeris, E., Kekos, D., Macris, B. & Chrsitakopoulos, P. (2003). Production and partial characterization of feruloyl esterase by *Sporotrichum thermophile* in solid-state fermentation. *Process Biochem* **38**, 1539-1543.

Tutino, M. L., Duilio, A., Parrilli, R., Remaut, E., Sannia, G. & Marino, G. (2001). A novel replication element from an Antarctic plasmid as a tool for the expression of proteins at low temperature. *Extremophiles* **5**, 257-64.

Uejima, A., Fukni, T., Fuknsaki, E., Omata, T., Kawamoto, T. & Tanaka, A. (1993). Efficient kinetic reslution of organosilicon compounds by stereoselective esterification with hydrolases in organic solvent. *Appl Microbiol Biotechnol* **38**, 426-486.

Valarini, P., Alvarez, M., Gasco, J., Guerrero, F. & Tokeshi, H. (2003). Assesment of soil properties by organic matter and em-microorganism incorporation. *Rev Bras Ci Solo* **27**, 519-525.

Van Kampen, M. & Egmond, M. (2000). Directed evolution: from a Staphylococcal lipase to a phospholipase. *Eur J Lipid Sci Technol* **102**, 717-7128.

Vath, G. M., Earhart, C. A., Monie, D. D., Iandolo, J. J., Schlievert, P. M. & Ohlendorf, D. H. (1999). The crystal structure of exfoliative toxin B: a superantigen with enzymatic activity. *Biochemistry* **38**, 10239-46.

Vieille, C. & Zeikus, G. J. (2001). Hyperthermophilic enzymes: sources, uses, and molecular mechanisms for thermostability. *Microbiol Mol Biol Rev* **65**, 1-43.

Vincent, D. & Lagreu, R. (1981). Identification od aspirinase with one of the carboxylesterase requiring a thiol group. *Biochem J* **197**, 771-773.

Volker, U., Andersen, K. K., Antelmann, H., Devine, K. M. & Hecker, M. (1998). One of two osmC homologs in Bacillus subtilis is part of the sigmaB-dependent general stress regulon. *J Bacteriol* **180**, 4212-8.

Wahler, D. & Reymond, J. L. (2001). High-throughput screening for biocatalysts. *Curr Opin Biotechnol* **12**, 535-44.

Wallon, G., Lovett, S. T., Magyar, C., Svingor, A., Szilagyi, A., Zavodszky, P., Ringe, D. & Petsko, G. A. (1997). Sequence and homology model of 3-isopropylmalate dehydrogenase from the psychrotrophic bacterium Vibrio sp. I5 suggest reasons for thermal instability. *Protein Eng* **10**, 665-72.

Wei, Y., Schottel, J. L., Derewenda, U., Swenson, L., Patkar, S. & Derewenda, Z. S. (1995). A novel variant of the catalytic triad in the Streptomyces scabies esterase. *Nat Struct Biol* **2**, 218-23.

Wezel, V., Vlaardingen, A. V., Posthumus, P., Crommentujin, R. & Sijm, G. (2000). Exposure routes of dibutyl-phthalate (DHP) and dietyl-hexyl-phthalate (DEHP). *Ecotoxicol Environ Saftey* **46**, 305-321.

Wheeler, G. E., Coleman, R. & Finean, J. B. (1972). Cholinesterase activities in subcellular fractions of rat liver. Association of acetylcholinesterase with the surface membrane and other properties of the enzyme. *Biochim Biophys Acta* **255**, 917-30.

Whitaker, J. R. (1972). Principles of enzymology for the food sciences. 481-501.

Willerslev, E., Hansen, A. J., Christensen, B., Steffensen, J. P. & Arctander, P. (1999). Diversity of Holocene life forms in fossil glacier ice. *Proc Natl Acad Sci U S A* **96**, 8017-21.

Winkler, U. K. & Stuckmann, M. (1979). Glycogen, hyaluronate, and some other polysaccharides greatly enhance the formation of exolipase by Serratia marcescens. *J Bacteriol* **138**, 663-70.

Wrasidlo, W., Schroder, U., Bernt, K., Hubener, N., Shabat, D., Gaedicke, G. & Lode, H. (2002). Synthesis, hydrolytic activation and cytotoxicity of etoposide prodrugs. *Bioorg Med Chem Lett* **12**, 557-60.

Zhou, X., Scharf, M. E., Sarath, G., Meinke, L. J., Chandler, L. D. & Siegfried, B. D. (2004). Partial purification and characterization of a methyl parathion resistance-associated general esterase in *Diabrotica virgifera* (Coleoptera: chrysomelidae). *Pesticide Biochem Physiol* **78**, 114-125.

Zock, J., Cantwell, C., Swartling, J., Hodges, R., Pohl, T., Sutton, K., Rosteck, P., Jr., Mcgilvray, D. & Queener, S. (1994). The Bacillus subtilis pnbA gene encoding p-nitrobenzyl esterase: cloning, sequence and high-level expression in Escherichia coli. *Gene* **151**, 37-43.

Acknowledgements

Four years have passed by since I first arrived at Germany. And here I am four years later holding a PHD degree. I have met a lot of people and each and every one of them affected my life in one way or another. Some of them made me laugh, some made me cry, others made me think, and some made me contemplate. And now while I am writing these lines, the picture of each and every one of you is passing in front of me, between the lines, behind the comas and on my screen. I met a great people indeed and learned a lot from it. "O mankind! Lo! We have created you male and female, and have made you nations and tribes that ye may know one another. Lo! the noblest of you, in the sight of Allah, is the best in conduct. Lo! Allah is Knower, Aware." Holy Quran, 049.013.

Special thanks are due to Prof. Antranikian, who gave me the opportunity to learn and do research at his institute and to Dr. Hashwa who supported me in every step I took to start my PHD work.
I would like to thank the DAAD (Deutscher Akademischer Austausch Dienst) who supported me financially and who did every thing possible and imaginable to make my arrival and stay in Germany perfect and comfortable.
I would like to thank Dr. Farah Qoura for all the help and advice he offered me during my stay in Germany and in the course of my research.
Special acknowledgments are due to Dana Biemann who always had a great smile on her face. "A smile costs nothing, but creates much. It enriches those who receive, without impoverishing those who give."
A special thank you to Dr. Martin Borchert for introducing me to Linux. For the time being I think Windows wins…
Many thanks are to Marina Royter, Moritz Katzer, Matthias Hess, Barbara Piela, Volker Thiemann, Imke Haller, Anke Peters, Ute Lorenz, Nele Stößer, Sandra Off, Sabina Riessen, Karen Sönnenberger, Daniel Wieber, Christiane Dock, Petra Esselun, Karna Benz and Ralf Grote for being the greatest colleagues ever.

I would like also to thank all my friends in Germany who showed me a great hospitality.

Finally, I would like to express my warmest love to my parents and family who always stood by me and never forgot me in their prayers.

Rami Al Khudary
Born in Tripoli - Lebanon 19.11.1977
Mobile # (Germany): +49 173 2035644
Mobile # (Lebanon): +961 3 110241
Tel. #   (Lebanon): +961 6 410296
Email: rami@khudary.com / khudary@gmail.com

Educational background	• October 2006: A PHD (Doctor rer. nat.) degree at the Technical University of Hamburg – Harburg (TUHH) "Institute of Technical Microbiology & Biotechnology"
	• June 2002: M.S. in Biology; *Lebanese American University* – Byblos (LAU)
	• July 1999: Teaching Diploma (for the Intermediate and Secondary classes); *American University of Beirut* (AUB)
	• July 1998: B.S. in Biology; *American University of Beirut* (AUB)
	• June 1995: Received the Lebanese Baccalaureate (II) at the *Tripoli Evangelical School for Girls & Boys* (TES) (Experimental Sciences Section)
Awards received	• A Prize given by the "Association of Lebanese Industrialists" for my work regarding "Biodegradation of Olive oil rich Wastewater under Aerobic & Mesophilic Conditions" "Lebanese Industrial Research Achievements" - LIRA - Fourth Conference & Exhibition.
	• Nabeel Mikary Prize for the top student in class. (TES)
	• Elias Abu Rustum Prize for the top student in the Exp. Sc. Class. (TES)
Scholarships	• A full scholarship granted by the "Deutscher Akademischer Austausch Dienst" – (German Academic Exchange Service - Bonn) – (DAAD) that covers my living and tuition expenses at TUHH.
	• A full scholarship granted by the "Hariri Foundation" that covered 4 years of tuition at (AUB)
Work experience	• 2003-2006: Gave several practical lab courses: From gene to product, Environmental microbiology and General microbiology. (TUHH)
	• 2001-2002: Research project coordinator: Zibar – Olive Mill Waste Water (OMW) mesophilic bio-treatment (L.A.U.)
	• 2000-2001: Gave Cell Biology and Botany lab courses. (L.A.U. –Byblos)
	• 2000-2001: Taught first, second and third secondary classes at Al Iman School (Science section) (Tripoli-Lebanon)
	• 1998-1999: Taught first, second and third secondary classes (Science and literary sections) at the Ahlieh School (Beirut- Lebanon)
Languages	• English and Arabic fluently spoken, excellently read and written. (TOEFL score: 287 - Computer based test). German well spoken, read and written. (Took a 6 month German Language course at Goethe Institute – Goettingen. DSH Prufung score: 80.8%)